Precision Poultry Farming

Precision Poultry Farming

Editors

Yang Zhao
María Cambra-López
Daniella Jorge De Moura
Weichao Zheng

MDPI • Basel • Beijing • Wuhan • Barcelona • Belgrade • Manchester • Tokyo • Cluj • Tianjin

Editors

Yang Zhao
The University of Tennessee
USA

María Cambra-López
Universitat Politècnica de València
Spain

Daniella Jorge De Moura
State University of Campinas
Brazil

Weichao Zheng
China Agricultural University
China

Editorial Office
MDPI
St. Alban-Anlage 66
4052 Basel, Switzerland

This is a reprint of articles from the Special Issue published online in the open access journal *Animals* (ISSN 2076-2615) (available at: https://www.mdpi.com/journal/animals/special_issues/precision_poultry_farming).

For citation purposes, cite each article independently as indicated on the article page online and as indicated below:

LastName, A.A.; LastName, B.B.; LastName, C.C. Article Title. *Journal Name* **Year**, *Volume Number*, Page Range.

ISBN 978-3-0365-4879-1 (Hbk)
ISBN 978-3-0365-4880-7 (PDF)

Contents

About the Editors

Yang Zhao

Dr. Yang Zhao is an Assistant Professor at Animal Science department at the University of Tennessee. Dr. Zhao received his B.S. and M.S. degrees from China Agricultural University and a PhD degree from Wageningen University in the Netherlands. Before joining the University of Tennessee, he was an assistant professor of Agricultural and Biological Engineering at Mississippi State University for three and half years. Dr. Zhao serves as the chair and member in several academic and award committees of the American Society of Agricultural and Biological Engineers (ASABE) and is an associate editor of the Transactions of the ASABE. His research focuses on smart poultry farming that addresses challenges in poultry production regarding sensors, automation, behavior monitoring, welfare assessment, disease prevention, and environment management. His work has been published in over 120 scientific articles. Dr. Zhao is the awardee of the ASABE Sunkist Young Designer Award, the AOC Early Career Award, and the Gamma Sigma Delta Research Award. Other recognitions of his work include ASABE Superior Paper Awards, Poultry Science Highly Cited Paper of the Year, Poultry Science Article of Editor's Choice, and ASABE Outstanding Reviewer.

María Cambra-López

Dr. María Cambra-López is Full Professor in the Animal Science Department at Universitat Politècnica de Valencia, UPV (Spain). She conducted her PhD in Agricultural Engineering at UPV (Spain) and completed her pre- and post-doc period at the Livestock Research Group at Wageningen University (The Netherlands). She has 17 years of research experience in the field of animal production and precision nutrition in non-ruminants. She has published 50 scientific papers (h-index=16), 1 book, and 3 book chapters. She has participated in 14 R&D projects funded in public calls and she has led 3 of them as the principal investigator. She has been awarded seven positive recognitions of her research activity (including the ASABE Superior Paper Award, American Society of Agricultural and Biological Engineers twice (2010 and 2012) and the PhD Extraordinary Prize at UPV). She is member of the European Association of Precision Livestock Farming (EA-PLF). Her professional career is characterized by an active participation in scientific forums, committees and reviews in journals. She currently coordinates the ANTS Research and Transfer service (Animal Nutrition and Technology Service, https://antsanimalnutrition.com), where she conducts research, development and innovation projects in animal production and nutrition.

Daniella Jorge De Moura

Dr. Daniella Jorge De Moura is an Associate Professor at the School of Agricultural Engineering at the State University of Campinas. Dr. Moura received her B.S. degree in Agronomy from University of São Paulo and M.S. and PhD degrees from State University of Campinas, both in Brazil. She attended a sandwich program at University of Florida during her PhD. Her research focuses on smart poultry and swine farming, including behavior monitoring, welfare assessment and environment management. In recent years, she coordinated several research projects, including researchers' and students' international mobility. As a result, she has published over 120 scientific articles and book chapters.

Weichao Zheng

Dr. Weichao Zheng is a Full Professor at the Department of Agricultural Structure and Bioenvironmental Engineering, China Agricultural University, China. Dr. Zheng received his B.S. and PhD degrees from China Agricultural University. Before joining China Agricultural University, he was a postdoctoral associate of Agricultural and Biological Engineering at the University of Illinois at Urbana-Champaign for two years. Dr. Zheng serves as the secretary-general of Animal Husbandry Engineering Commission of the Chinese Society of Agricultural Engineering. His research focuses on new housing systems for welfare, precision environment control, and airborne disease prevention for poultry. He has been the PI of six national and provincial-level research projects. Recognitions of his research contributions include 6 provincial-level science and technology awards, over 70 published journal articles, and 15 Chinese national patents.

Preface to "Precision Poultry Farming"

The increase in the global population comes along with growing demands on protein resources. To meet such demands, the global production of poultry meat and eggs has skyrocketed in the past few decades and is projected to continue growing in decades to come. While poultry production makes crucial contributions to food and nutrition security, it uses substantial natural and human resources and has significant impacts on society, the economy, public health, and the environment. Although the extent of these impacts may vary among continents and countries due to differences in production practices and social structures and preferences, the global poultry industry, as a whole, should strive to keep improving sustainability and efficiency in its resource usage. Precision Poultry Farming (PPF) features applications of continuous, objective, and automated sensing technologies and computer tools for sustainable and efficient poultry production; it offers the poultry industry solutions to address challenges in terms of poultry management, the environment, nutrition, automation and robotics, health, welfare assessment, behavior monitoring, waste management, etc.

This Special Issue on "Precision Poultry Farming" was initiated in early 2020 by the Guest Editors with assistance from the Managing Editor of /Animals/. A total 12 high quality manuscripts were eventually published, covering a wide spectrum of the most recent research on PPF. The contributing authors come from Asia, Europe, North America, and South America. We appreciate all authors, as well as reviewers, for their contributions to this successful collection.

Yang Zhao, María Cambra-López, Daniella Jorge De Moura, and Weichao Zheng
Editors

animals

MDPI

Article

Accuracy of Broiler Activity Index as Affected by Sampling Time Interval

Xiao Yang [1], Yang Zhao [1,*] and George T. Tabler [2]

1 Department of Agricultural and Biological Engineering, Mississippi State University, Mississippi, MS 39762, USA; xy123@msstate.edu
2 Department of Poultry Science, Mississippi State University, Mississippi, MS 39762, USA; ttabler@poultry.msstate.edu
* Correspondence: yzhao@abe.msstate.edu

Received: 29 May 2020; Accepted: 25 June 2020; Published: 26 June 2020

Simple Summary: Broiler activity index is a measure of bird movement through determining bird-representative pixel changes between consecutive images. Since the concept of activity index was coined, different sampling time intervals of consecutive images have been used to determine broiler activity. In this study, we found that accuracy of broiler activity decreased at longer sampling time intervals, with the 0.04-s interval yielding the most accurate activity index among all intervals investigated. In addition, broiler activity in the commercial house generally decreased as birds aged and varied at different monitoring locations. The research provides insights into image-sampling strategies for accurately determining broiler activity index, which may help to address growing public concerns on poultry welfare and health.

Abstract: Different time intervals between consecutive images have been used to determine broiler activity index (AI). However, the accuracy of broiler AI as affected by sampling time interval remains to be explored. The objective of this study was to investigate the effect of the sampling time interval (0.04, 0.2, 1, 10, 60, and 300 s) on the accuracy of broiler AI at different bird ages (1–7 weeks), locations (feeder, drinker, and open areas) and times of day (06:00–07:00 h, 12:00–13:00 h, and 18:00–19:00 h). A ceiling-mounted camera was used to capture top-view videos for broiler AI calculations. The results show that the sampling time interval of 0.04 s yielded the highest broiler AI because more bird motion details were captured at this short time interval. The broiler AIs at longer time intervals were 1–99% of that determined at the 0.04-s interval. The broiler AI at 0.2-s interval showed an acceptable accuracy with 80% less computational resources. Broiler AI decreased as birds aged but increased after week 4 at the drinker area. Broiler AI was the highest at the open area for weeks 1–4 and at the feeder and drinker areas for weeks 5–7. It is concluded that the accuracy of broiler AI was significantly affected by sampling time intervals. Broiler AI in commercial housing showed both temporal and spatial variations.

Keywords: broiler; activity index; time interval; age; image processing

1. Introduction

Broiler activity is considered a major indicator of animal physical and physiological conditions [1,2]. It was reported by Thorp and Duff [3] that exercising broilers a few times every day could benefit broilers' leg skeletal conditions [4], thus reducing the incidence of lameness and improving bird walking ability [5]. More movement and activity by broilers may also help to reduce the prevalence of hock burns [6], footpad dermatitis [7], and breast burns [8] by reducing the contact with wet litter.

In order to quantify the animal activity, activity index (AI), a measure of movement intensity through image processing, was proposed by Bloemen et al. [9]. Activity index was defined as the percentage of pixels of moving animals to the total number of pixels within the image (including animals and background). In more recent research, the total number of pixels was replaced with total bird-representative pixels to compensate for variations in animal size at different ages [10–12]. Since the concept was coined, AI has been widely used to quantify the activities of broilers [13,14].

Broiler AI is calculated by determining changes in bird-representative pixels between consecutive images. Different from the optical flow statistics that Dawkin et al. [15] used to derive measures of broiler behaviors and gaits in commercial farm, the method of AI in this study only considers the amount of movement between consecutive images while the movement direction is not included. Using a short time interval between consecutive images may capture more movement details and yield better AI accuracy; however, a time interval that is unnecessarily short (e.g., yielding too many consecutive images for only trivial bird movements) cannot further improve AI accuracy and may increase image processing time. Longer time intervals, on the other hand, may miss identifying birds' movements and compromise AI accuracy. Therefore, a proper time interval is important for ensuring the accuracy of AI while improving processing efficiency and saving computational resources. Different time intervals have been used to determine broiler AI in previous research. For instance, Neves et al. [14] analyzed the images at 60 s intervals to determine bird activity as affected by feeder types. Bloemen et al. [9] sampled images with a time interval of 5 s to investigate the effect of the thermal environment on broiler activity. Aydin et al. [10] used a time interval of 0.2 s to measure the activity of broilers with different gait scores. However, proper time intervals remain to be explored.

Selection of a proper time interval should be based on the birds' movement intensity, which could be affected by many factors, such as bird age, location, and time of day. Bird movement intensity may vary with bird age due to changes in their physical conditions, like body weight and walking ability [11,16]. Broilers do not always spread out evenly on the floor areas in commercial houses, resulting in different degrees of crowdedness that may affect bird movements. For example, Arnould et al. [17] found that areas near feeders and drinkers were more crowded because broilers tended to stay and rest near the sources of feed and water. Furthermore, bird movement intensity varies within a day. Previous research showed that birds are more active during the first hour after light ON and before light OFF [18]. However, the effects of the above-mentioned temporal and spatial variations in broiler movement intensity on broiler AI have not been investigated.

The objective of this study was to investigate the effect of sampling time interval (0.04, 0.2, 1, 10, 60, and 300 s) on the accuracy of broiler AI. The effects of bird age (1–7 weeks), location (feeder, drinker, and open area) and time of day (06:00–07:00 h, 12:00–13:00 h and 18:00–19:00 h) on broiler AI were also examined.

2. Materials and Methods

2.1. Housing, Animals and Management

The study was conducted in a commercial broiler house located at Mississippi State University during 12/2019–1/2020. The house measured 120 × 13 × 3 m (L × W × H) with a capacity of 16,120 Ross 708 straight run broilers and a production cycle of 8 weeks. All chicks were purchased from a commercial poultry hatchery in Mississippi. Both tray and tube feeders were used in weeks 1 and 2 of bird age, then tray feeders were removed from week 3. Flock management and diets followed the typical procedures in the industry. The lighting schedule was set to 24L:0D from 1 d to 7 d, 20L:4D from 8 d to 56 d. The light intensity was set to 54 lux from 1d to 13 d, then gradually dimmed to 3 lux by 20 d and kept at 3 lux till 56 d. Lights were turned on at 05:00 h and turned off at 01:00 h of the next day.

2.2. Camera System

A fisheye IP camera (Dahua, IPC-EW4431-ASW, Dahua Technology USA Inc., Irvine, CA, USA) was installed on the ceiling (height = 3 m), located in the middle of the house. The frame rate of the camera was 25 frames per second. Three one-hour video clips (06:00 to 07:00 h, 12:00 to 13:00 h and 18:00 to 19:00 h) were recorded on Wednesday every week. The video clips were converted into images with all frames extracted (time interval of 0.04 s between consecutive images) or partial frames at time intervals of 0.2, 1, 10, 60 and 300 s. Images were firstly corrected for distortions using Python (Python 3.7.1, Python Software Foundation, Beaverton, OR, USA). Afterwards, three specific areas (200 × 200 pixels) located at the feeder, drinker and open area (Figure 1) were cropped out of the images and fed to MATLAB (2018b, The MathWorks, Inc. Natick, MA, USA) for image processing. The dimension of cropped area was equivalent to the actual area of 0.71 × 0.71 m. The images were used to calculate AI at the drinker and open areas in weeks 1–7, and at feeding area in weeks 3–7 when only tube feeders were provided.

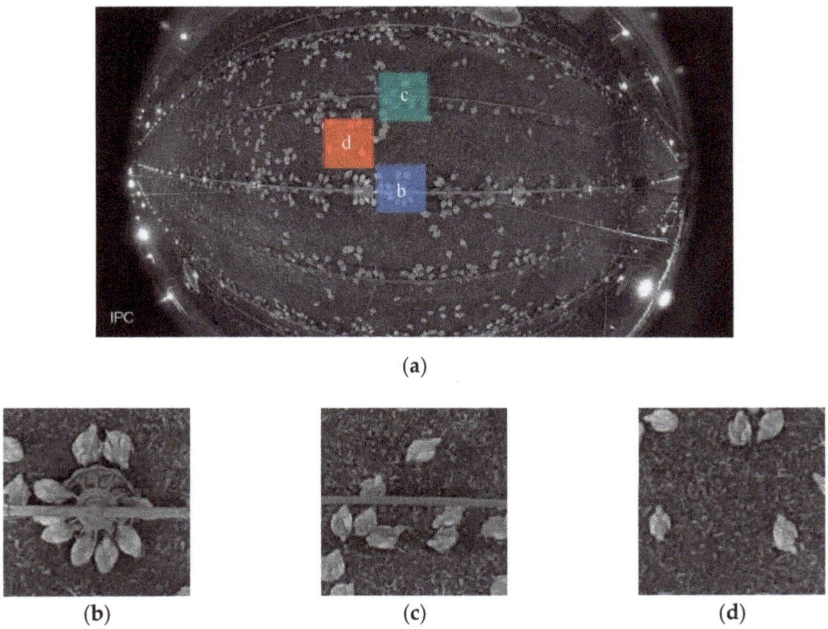

Figure 1. Example images of (**a**) original image (**b**) feeder (**c**) drinker and (**d**) open area.

2.3. Activity Index

The value $I(x, y)$ represents the intensity of the pixel at coordinates (x, y) in that image. The difference in intensity between the current image $I(x, y, t)$ and the previous image $I(x, y, t-1)$, was calculated by subtraction of the two consecutive images. A resulting image of $I_a(x, y, t)$, in binary form, was then generated according to results of the subtraction (Equation (1)).

$$I_a(x,y,t) = \begin{cases} 1, & if\ I(x,y,t) - I(x,y,t-1) > \tau \\ 0, & otherwise \end{cases} \quad (1)$$

The threshold (τ) was set to 15% of the maximal intensity of each video clip by observing the "empty" background of first 20 frames for each clip to avoid erroneous results due to noise, e.g., electrical noise in the coaxial cabling and image acquisition circuits, and lighting variations.

Total number of non-zero pixels was calculated as the variation between the two consecutive images due to the activity of broilers. To compensate for the size and number of the birds, the activity index AI(t) was calculated as the fraction of the number of non-zero pixels in the resulting binary image I_a(x, y, t) with respect to the number of broiler-representative pixels S(t − 1) in the previous image I(x, y, t − 1) (Equation (2)).

$$AI(t) = \frac{\sum I_a(x, y, t)}{S(t-1)} \quad (2)$$

2.4. Data Preparation and Statistic Analysis

The effects of sampling time interval, bird age, sampling location, and time of the day, as well as the major two-way interactions, on broiler AI was analyzed using the PROC GLM (generalized linear model) procedure in SAS 10.9 (SAS Institute., Cary, NC, USA). A significant difference in multiple comparisons of group means was defined as $p < 0.05$. The levels for time intervals were 0.04, 0.2, 1, 10, 60 and 300 s. The levels of age were 1–7 weeks in drinker and open areas and 3–7 weeks in the feeder area. The levels of sampling location factor consisted of feeder, drinker and open areas. The levels of sampling time within a day were 06:00–07:00 h, 12:00–13:00 h and 18:00–19:00 h. In order to compare the difference of AI among different time intervals, the cumulative AI of every 300 s was calculated by simply adding the AIs within a 300-s duration. Some of the video clips were not strictly an hour long (56 to 59 min); therefore, 11 total samples were obtained from each video clip. For the effects of bird age, location and time of day, only data with time intervals of 0.04 s (full frames) were used for analysis.

3. Results

3.1. Time Interval

Table 1 shows the average broiler AI with different time intervals at different locations. A time interval of 0.04 s yielded the highest AI. With an increase in time interval, the broiler AI deceased from 100% (0.2 s) to 2% (300 s) of the AI determined with a time interval of 0.04 s ($p < 0.0001$ for all). No differences in AI were observed between the ratio of 0.04 s and 0.2 s at the feeder and drinker areas. However, a lower ratio of AI at the 0.2-s interval was found at the feeder area compared with the 0.04-s interval (($p < 0.0001$). Lower broiler AIs were observed at the 1-s time interval than at the 0.2-s interval at the feeder ($p < 0.0001$) and drinker ($p = 0.0267$) areas. At the open area, the broiler AI with a time interval of 1 s was lower than 0.04 s ($p = 0.0007$); however, no significant difference was observed between 0.2 s and 1 s.

Table 1. Average broiler activity index (AI) with different time intervals at different locations.

Time Interval (s)	Location					
	Feeder		Drinker		Open Area	
	AI	Ratio (%)	AI	Ratio (%)	AI	Ratio (%)
0.04	38.8 [a]	100	42.5 [a]	100	81.2 [a]	100
0.2	38.5 [a]	100	41.7 [a]	100	58.0 [ab]	84
1	27.3 [b]	74	31.1 [b]	92	34.3 [bc]	56
10	8.0 [c]	21	11.7 [c]	32	20.1 [cd]	27
60	2.3 [cd]	6	4.9 [cd]	12	10.2 [cd]	13
300	0.6 [d]	1	0.9 [d]	2	1.7 [d]	2
SEM	3.1	2	5.4	4	13.4	3

Ratio: AI calculated at a time interval relative to that at the 0.04-s time interval. SEM: pooled standard error mean for the main effects of location. [a,b,c,d] Means in the same column with different superscripts are different ($p < 0.05$). Five weeks (week 3–7) of data in the feeder area and seven weeks (week 1–7) of data in the drinker and open areas are summarized.

3.2. Age and Location

Table 2 shows the weekly average broiler AI at different locations. At the feeder area, the broiler AI in week 3 was higher than that in weeks 4 and 7 ($p < 0.0001$). No significant differences were observed in weeks 4, 6 and 7. At the age of week 5, the AI was higher than in weeks 4 ($p = 0.0024$) and 6 ($p = 0.0218$), however, not different from week 7. Broiler AI at the open area gradually decreased from week 1 to week 5, and no differences were found in weeks 4, 5, 6 and 7. No difference in broiler AI at the open area was observed between weeks 2 and 3. The AIs in weeks 4–7 were lower than those in weeks 1–3 ($p < 0.05$). For the effects of sampling locations, the broiler AIs at the open area were higher than at the drinker area in weeks 1 and 2 ($p = 0.0098$ and $p = 0.0011$, respectively). In weeks 3 and 4, the highest broiler AI was observed at the open area and the lowest AIs at the drinker area ($p < 0.0001$). In weeks 6 and 7, the broiler AIs at the feeder ($p = 0.0011$ and $p = 0.0134$, respectively) and drinker ($p = 0.0004$ and $p < 0.0001$, respectively) areas were higher than at the open area. Generally, broiler AI decreased as broilers aged up in all locations in this study.

Table 2. Weekly average broiler activity index (AI) at different locations.

Bird Age (Week)	Location			
	Feeder	Drinker	Open	SEM [1]
1	–	83.2^{Ba}	232.5^{Aa}	39.7
2	–	74.1^{Ba}	137.2^{Ab}	13.0
3	64.5^{Ba}	28.9^{Cbc}	95.8^{Ab}	7.2
4	28.6^{Ac}	16.5^{Bc}	32.6^{Ac}	1.9
5	37.8^{Ab}	20.7^{Bc}	21.6^{Bc}	1.7
6	30.9^{Ac}	31.5^{Abc}	23.1^{Bc}	1.6
7	31.9^{Bbc}	42.4^{Ab}	25.5^{Cc}	1.8
SEM [2]	2.1	6.1	14.8	–

Data with a time interval of 0.04 s were included. [1] SEM: Pooled standard error mean for age effect. [2] SEM: Pooled standard error mean for location effect. ^{A,B,C} Means in the same row with different superscripts are different ($p < 0.05$). ^{a,b,c} Means in the same column with different superscripts are different ($p < 0.05$).

3.3. *Time of Day*

Table 3 shows the weekly average broiler AI within three time periods at three locations. At the feeder area, differences among time periods were observed at 5 and 6 weeks of bird age. At the drinker area, higher broiler AIs were observed at either 12:00 h or 18:00 h during 1–7 weeks of bird age. The lowest broiler AIs were observed at 06:00 h, except in week 3 at the drinker area. At the open area, differences were found in weeks 4 and 6, with the highest AIs being identified at 12:00 h and the lowest at 06:00 h ($p = 0.0007$ and $p = 0.0026$, respectively).

Table 3. Weekly average broiler activity index (AI) within three time periods (06:00–07:00 h, 12:00–13:00 h, and 18:00–19:00 h) at three locations (feeder, drinker, and open area).

Bird Age (Week)	Location											
	Feeder				Drinker				Open			
	06:00	12:00	18:00	SEM	06:00	12:00	18:00	SEM	06:00	12:00	18:00	SEM
1	–	–	–	–	43.7^{B}	136.3^{A}	69.5^{B}	21.0	179.C	384.9	133.7	91.0
2	–	–	–	–	46.0^{B}	97.6^{A}	78.7^{AB}	11.4	91.7	159.0	160.7	28.3
3	60.2	64.0	69.2	6.4	31.6^{A}	35.9^{A}	19.1^{B}	3.1	98.8	85.9	102.6	20.8
4	24.3	30.5	31.1	2.6	6.1^{C}	19.6^{B}	23.8^{A}	1.3	22.5^{B}	42.3^{A}	32.9^{AB}	3.7
5	43.4^{A}	31.7^{B}	38.3^{AB}	2.3	18.7	19.1	24.2	3.1	18.5	26.3	20.0	3.2
6	28.8^{B}	28.2^{B}	35.7^{A}	2.2	25.1^{B}	31.8^{AB}	37.8^{A}	2.9	17.3^{3}	29.1^{A}	22.8^{AB}	2.5
7	28.9	32.2	34.6	2.8	41.3	44.2	41.7	3.2	28.7	23.6	24.1	3.5

SEM: Pooled standard error mean for main effects of time periods at each location. ^{A,B,C} Means in the same row with different superscripts under the same location category are different ($p < 0.05$). Data with a time interval of 0.04 s were used.

3.4. *Selection for Proper Time Interval*

Figure 2 shows the weekly average broiler AI at the feeder, drinker and open areas with different time intervals. At the feeder area, no difference in broiler AI was observed between the time intervals of 0.04 s and 0.2 s in week 3 or weeks 5–7. In week 4, the AI with the time interval of 0.2 s was higher than with 0.04 s ($p = 0.0320$). At the drinker area, the difference in broiler AI in weeks 1–4 and week 6 was not significant between 0.04-s and 0.2-s intervals. In week 5, the AI with the time interval of 1 s was higher than that with 0.04 s ($p = 0.0436$). At week 7, a time interval of 0.2 s yielded a lower AI than 0.04 s ($p = 0.0068$). At the open area, no differences in broiler AI between time intervals of 0.04 and 0.2 s were found in weeks 1–2 or weeks 4–6. In weeks 3 and 7, the AI with a time interval of 0.2 s was higher than for 0.04 s ($p = 0.0056$ and $p = 0.0159$).

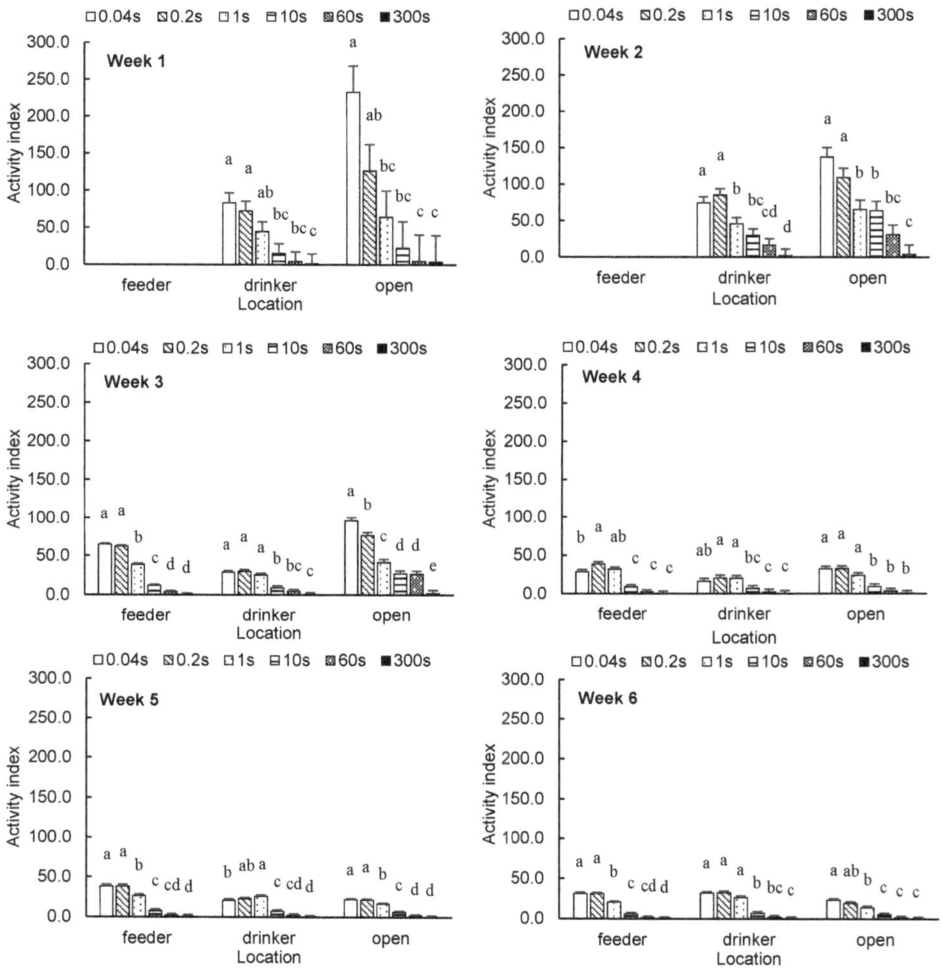

Figure 2. *Cont.*

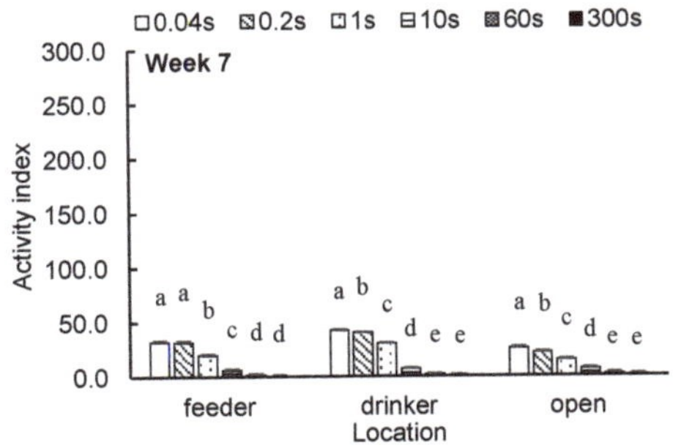

Figure 2. Weekly average broiler activity index (least square means ± SEM) at feeder, drinker and open areas with different time intervals. a,b,c,d,e Means with different superscripts within the same category are different ($p < 0.05$).

4. Discussion

Different sampling time intervals, from seconds to minutes, have been adopted to calculate the AI of livestock and poultry [10,12,19,20]. However, how the time intervals affect the accuracy of animal AI remains to be understood. In this study, we found that AI decreased at longer sampling time intervals (Table 1). This is because some transient behaviors, such as turning, bobbling, preening and shaking, can be miss-identified as the sampling time interval increases. Any time interval that lasts longer than the time of the transient behavior may fail to capture the information. Our results also show that a 0.2-s interval can be considered as an alternative to 0.04 s in broiler AI determination, because it delivered comparably high AI while reducing the image processing workload by 80%. Selection of proper sampling time intervals for accurate AI measurement of certain animal species should consider the velocities of both continuous movements (e.g., walking and running) and discrete movements (e.g., pecking, touching, smelling, dashing, etc.). A sampling time interval of 0.03 s has been used for cows [19,21], 0.04 s for pigs [20,22], and 0.2 s and 0.3 s for poultry [10,12,23].

Broiler AIs generally decreased as the birds got older (Table 2). This result is consistent with those previously reported by Alvino et al. [24] and Bizeray et al. [25], who also found older broilers were less active while gaining body weight [26]. Increases in AIs at the drinker and feeder areas were noticed after week 4 of bird age (Table 2). This is possibly because of birds' increasing demand for feed and water [27] as birds get older and heavier, promoting more traffic around drinkers and feeders.

Our results show obvious spatial variations in broiler AI, which can be explained by the different movement intensities of predominant behaviors in the locations of concern (i.e., feeder, drinker, and open areas). Feeding and drinking behaviors at feeder and drinker areas [28,29] do not involve more intensive body movements compared to walking, chasing, and playing behaviors (sparring, frolicking and food-running) [30] at open litter areas. Therefore, the broiler AIs in feeder and drinker areas were lower than that in open areas in weeks 1–4 (Table 2). As broilers got older and heavier, they became less active and tended to use the open area as a place for resting, which involves less body movement than feeding and drinking. As such, we found a lower broiler AI in open areas, as compared to feeder and drinker areas, in weeks 5–7.

Understanding the broiler AI at the feeder area may also help to clarify the birds' health status. For instance, Weeks et al. [31] compared the feeding patterns of broilers with different gait scores and found those birds with substantially impaired walking ability halved the number of feeding bouts but doubled the feeding time per visit to reach the same amount of feed intake as healthy birds. It should be noted that the AI in our study quantifies the bird activity within a 0.71 × 0.71 m area (including both the feeder/drinker and surrounding areas); thus, it involves not only the feeding and drinking behaviors but also the traffic around feeders and drinker lines. This is different from other measures of feed and water resource usage by broilers, such as feed intake [32] time budget of broiler feeding and drinking behaviors [33,34], and feeding and drinking frequency (times of feeder or drinker visiting within an hour) [35,36].

There were also temporal variations in broiler activity within a day (Table 3). The possible explanation for higher broiler AI at the feeder area in the early morning or afternoon could be the broiler diurnal rhythms. Li et al. [37] explored the effect of light intensity and spectrum on broiler (35–40 d) feeding behaviors with a schedule of 16L:8D and found the peak feeding occurred at 2–3 h after light ON and before light OFF. The high level of feeding behavior in the early morning could be a feed compensation after the dark period without any feed intake [38]. More feeding behavior at the end of a day was probably because the birds can anticipate the darkness [39] and spontaneously increase the feed intake before the light is turned OFF [28]. It should be pointed out that the broiler AI at the feeder area during 06:00–07:00 h did cover the early light ON period but was not consistently higher than that of the midday period. The discrepancy is possibly because of the difference in bird age or lighting schedule [40]. In addition, the movement of birds around the feeder, which was equivalent to the movement of birds at the open area, could be another reason for higher AI during 12:00–13:00 h at the feeder area. There was no clear temporal pattern in AI at the drinker area based on our results. Schwean-Lardner et al. [41] studied the effect of daytime length on broiler (27 d and 42 d) behavioral patterns and found a similar rhythm of drinking behavior as of feeding. In another study that explored the effect of feeding and lighting programs on broiler feeding and drinking patterns indicated that the drinking patterns of broilers were largely influenced by lighting programs but independent of feeding behaviors [40]. One explanation for the discrepancy could be the lighting schedule. Another reason could be the movement of birds near the drinker line within the concerned area. As for open area, higher broiler AI was typically found at middle of the day in our study, which is consistent with the results reported by Sherlock et al. [42].

It worth noting that the AI is a measure of bird movements in two dimensions on a horizontal plane. Bird movements in the vertical direction for behaviors such as standing up, lying down, and pecking the feeder and drinker are not included in the AI calculation, but may have important implications for poultry welfare, health, and production efficiency. To account for the birds' vertical motion, cameras with a depth sensor or body-mounted sensors that produce 3D motion measurements may be considered [12].

5. Conclusions

In this study, the effects of sampling time interval between consecutive images on the accuracy of broiler AI at different bird ages, sampling locations and times of day were investigated using image processing. We conclude that a sampling time interval of 0.04 s yielded the best broiler AI results. The AI results for a 0.2-s time interval were acceptable but required significantly less computational resource usage. At different ages, broiler AIs at the feeder and open area generally decreased from week 1 to week 7. However, an increase at the drinker area was observed after week 4. For the effects of location, higher AIs during weeks 1–4 occurred at the open areas, and this switched to feeder and drinker areas during weeks 5–7. Clear diurnal behavioral rhythms were also found at feeder and open areas. In summary, broiler AIs in commercial housing showed both temporal and spatial variations. These findings provide important insights into accurate broiler activity measurement for broiler welfare, health, and production evaluation.

Author Contributions: Conceptualization, X.Y. and Y.Z.; data curation, X.Y.; formal analysis, X.Y.; funding acquisition, Y.Z.; investigation, X.Y. and Y.Z.; project administration, Y.Z. and G.T.T.; methodology, X.Y. and Y.Z.; supervision, Y.Z. and G.T.T.; resources, Y.Z. and G.T.T.; writing—original draft, X.Y.; writing—review and editing, Y.Z. and G.T.T. All authors have read and agreed to the published version of the manuscript.

Funding: This study was financially supported by the Foundation for Food and Agriculture Research (FFAR) SMART Broiler Initiative and the Mississippi Agricultural and Forestry Experiment Station (MAFES) Special Research Initiative.

Acknowledgments: The authors appreciate the support and assistance provided by the farm staff at Mississippi State University.

Conflicts of Interest: The authors declare no conflict of interest. The funders had no role in the design of the study; in the collection, analyses, or interpretation of data; in the writing of the manuscript; or in the decision to publish the results.

References

1. Bizeray, D.; Estevez, I.; Leterrier, C.; Faure, J.M. Influence of increased environmental complexity on leg condition, performance, and level of fearfulness in broilers. *Poult. Sci.* **2002**, *81*, 767–773. [CrossRef]
2. Simsek, U.G.; Dalkilic, B.; Ciftci, M.; Cerci, I.H.; Bahşi, M. Effects of Enriched Housing Design on Broiler Performance, Welfare, Chicken Meat Composition and Serum Cholesterol. *Acta Veter. Brno* **2009**, *78*, 67–74. [CrossRef]
3. Thorp, B.; Duff, S. Effect of exercise on the vascular pattern in the bone extremities of broiler fowl. *Res. Veter. Sci.* **1988**, *45*, 72–77. [CrossRef]
4. Hester, P.Y. The Role of Environment and Management on Leg Abnormalities in Meat-Type Fowl. *Poult. Sci.* **1994**, *73*, 904–915. [CrossRef] [PubMed]
5. Wilson, J.L.; Weaver, W.D.; Beane, W.L.; Cherry, J.A. Effects of Light and Feeding Space on Leg Abnormalities in Broilers. *Poult. Sci.* **1984**, *63*, 565–567. [CrossRef] [PubMed]
6. Kaukonen, E.; Norring, M.; Valros, A. Effect of litter quality on foot pad dermatitis, hock burns and breast blisters in broiler breeders during the production period. *Avian Pathol.* **2016**, *45*, 667–673. [CrossRef] [PubMed]
7. Shepherd, E.; Fairchild, B. Footpad Dermatitis in Poultry. *Poult. Sci.* **2010**, *89*, 2043–2051. [CrossRef]
8. Haslam, S.; Knowles, T.; Brown, S.; Wilkins, L.; Kestin, S.; Warriss, P.; Nicol, C. Factors affecting the prevalence of foot pad dermatitis, hock burn and breast burn in broiler chicken. *Br. Poult. Sci.* **2007**, *48*, 264–275. [CrossRef]
9. Bloemen, H.; Aerts, J.M.; Berckmans, D.; Goedseels, V. Image analysis to measure activity index of animals. *Equine Veter. J.* **2010**, *29*, 16–19. [CrossRef]
10. Aydin, A.; Cangar, Ö.; Ozcan, S.E.; Bähr, C.; Berckmans, D. Application of a fully automatic analysis tool to assess the activity of broiler chickens with different gait scores. *Comput. Electron. Agric.* **2010**, *73*, 194–199. [CrossRef]
11. Silvera, A.M.; Knowles, T.G.; Butterworth, A.; Berckmans, D.; Vranken, E.; Blokhuis, H. Lameness assessment with automatic monitoring of activity in commercial broiler flocks. *Poult. Sci.* **2017**, *96*, 2013–2017. [CrossRef] [PubMed]
12. Yang, X.; Huo, X.; Li, G.; Purswell, J.L.; Tabler, T.; Chesser, D.; Zhao, Y. Application of Elevated Perching Platform and Robotic Vehicle in Broiler Production. In Proceedings of the ASABE Annual International Meeting, Boston, MA, USA, 7–10 July 2019. [CrossRef]
13. Kristensen, H.; Aerts, J.; Leroy, T.; Wathes, C.; Berckmans, D. Modelling the dynamic activity of broiler chickens in response to step-wise changes in light intensity. *Appl. Anim. Behav. Sci.* **2006**, *101*, 125–143. [CrossRef]
14. Neves, D.P.; Mehdizadeh, S.A.; Tscharke, M.; de Alencar Nääs, I.M.; Banhazi, T.M. Detection of flock movement and behaviour of broiler chickens at different feeders using image analysis. *Inf. Process. Agric.* **2015**, *2*, 177–182. [CrossRef]
15. Dawkins, M.S.; Lee, H.-J.; Waitt, C.D.; Roberts, S.J. Optical flow patterns in broiler chicken flocks as automated measures of behaviour and gait. *Appl. Anim. Behav. Sci.* **2009**, *119*, 203–209. [CrossRef]
16. Bessei, W. Welfare of broilers: A review. *World's Poult. Sci. J.* **2006**, *62*, 455–466. [CrossRef]
17. Arnould, C.; Faure, J.M. Use of pen space and activity of broiler chickens reared at two different densities. *Appl. Anim. Behav. Sci.* **2003**, *84*, 281–296. [CrossRef]
18. Kristensen, H.H.; Cornou, C. Automatic detection of deviations in activity levels in groups of broiler chickens—A pilot study. *Biosyst. Eng.* **2011**, *109*, 369–376. [CrossRef]
19. Poursaberi, A.; Bähr, C.; Pluk, A.; Van Nuffel, A.; Berckmans, D. Real-time automatic lameness detection based on back posture extraction in dairy cattle: Shape analysis of cow with image processing techniques. *Comput. Electron. Agric.* **2010**, *74*, 110–119. [CrossRef]
20. Kashiha, M.A.; Bahr, C.; Ott, S.; Moons, C.P.; Niewold, T.; Tuyttens, F.A.; Berckmans, D. Automatic monitoring of pig locomotion using image analysis. *Livest. Sci.* **2014**, *159*, 141–148. [CrossRef]

21. Chapinal, N.; De Passillé, A.; Weary, D.; Von Keyserlingk, M.; Rushen, J. Using gait score, walking speed, and lying behavior to detect hoof lesions in dairy cows. *J. Dairy Sci.* **2009**, *92*, 4365–4374. [CrossRef]
22. Von Wachenfelt, H.; Pinzke, S.; Nilsson, C. Gait and force analysis of provoked pig gait on clean and fouled concrete surfaces. *Biosyst. Eng.* **2009**, *104*, 534–544. [CrossRef]
23. Tickle, P.G.; Hutchinson, J.R.; Codd, J. Energy allocation and behaviour in the growing broiler chicken. *Sci. Rep.* **2018**, *8*, 4562. [CrossRef]
24. Alvino, G.; Archer, G.; Mench, J. Behavioural time budgets of broiler chickens reared in varying light intensities. *Appl. Anim. Behav. Sci.* **2009**, *118*, 54–61. [CrossRef]
25. Bizeray, D.; Leterrier, C.; Constantin, P.; Picard, M.; Faure, J. Early locomotor behaviour in genetic stocks of chickens with different growth rates. *Appl. Anim. Behav. Sci.* **2000**, *68*, 231–242. [CrossRef]
26. EFSA. Scientific Opinion on the influence of genetic parameters on the welfare and the resistance to stress of commercial broilers. *EFSA J.* **2010**, *8*, 1666. [CrossRef]
27. Bruno, L.; Maiorka, A.; Macari, M.; Furlan, R.; Givisiez, P. Water intake behavior of broiler chickens exposed to heat stress and drinking from bell or and nipple drinkers. *Rev. Bras. Ciênc. Avíc.* **2011**, *13*, 147–152. [CrossRef]
28. Kristensen, H.H.; Prescott, N.B.; Perry, G.C.; Ladewig, J.; Ersbøll, A.K.; Overvad, K.C.; Wathes, C.M. The behaviour of broiler chickens in different light sources and illuminances. *Appl. Anim. Behav. Sci.* **2007**, *103*, 75–89. [CrossRef]
29. Svihus, B.; Lund, V.; Borjgen, B.; Bedford, M.; Bakken, M. Effect of intermittent feeding, structural components and phytase on performance and behaviour of broiler chickens. *Br. Poult. Sci.* **2013**, *54*, 222–230. [CrossRef]
30. Baxter, M.; Bailie, C.L.; O'Connell, N. Play behaviour, fear responses and activity levels in commercial broiler chickens provided with preferred environmental enrichments. *Animal* **2018**, *13*, 171–179. [CrossRef]
31. Weeks, C.; Danbury, T.; Davies, H.; Hunt, P.; Kestin, S. The behaviour of broiler chickens and its modification by lameness. *Appl. Anim. Behav. Sci.* **2000**, *67*, 111–125. [CrossRef]
32. Bai, S.P.; Wu, A.M.; Ding, X.M.; Lei, Y.; Bai, J.; Zhang, K.Y.; Chio, J.S. Effects of probiotic-supplemented diets on growth performance and intestinal immune characteristics of broiler chickens. *Poult. Sci.* **2013**, *92*, 663–670. [CrossRef] [PubMed]
33. Sultana, S.; Hassan, R.; Choe, H.S.; Ryu, K.S. The Effect of Monochromatic and Mixed LED Light Colour on the Behaviour and Fear Responses of Broiler Chicken. *Avian Boil. Res.* **2013**, *6*, 207–214. [CrossRef]
34. Ventura, B.A.; Siewerdt, F.; Estevez, I. Access to Barrier Perches Improves Behavior Repertoire in Broilers. *PLoS ONE* **2012**, *7*, e29826. [CrossRef]
35. Hocking, P.M.; Rutherford, K.M.D.; Picard, M. Comparison of time-based frequencies, fractal analysis and T-patterns for assessing behavioural changes in broiler breeders fed on two diets at two levels of feed restriction: A case study. *Appl. Anim. Behav. Sci.* **2007**, *104*, 37–48. [CrossRef]
36. Bizeray, D.; Estevez, I.; Leterrier, C.; Faure, J. Effects of increasing environmental complexity on the physical activity of broiler chickens. *Appl. Anim. Behav. Sci.* **2002**, *79*, 27–41. [CrossRef]
37. Li, G.; Zhao, Y.; Purswell, J.L.; Liang, Y.; Lowe, J.W. Feeding Behaviors of Broilers at Chicken-perceived vs. Human-perceived Light Intensities under Two Light Spectrums. In Proceedings of the ASABE Annual International Meeting, Detroit, MI, USA, 29 July–1 August 2018. [CrossRef]
38. Rodrigues, I.; Choct, M. Feed intake pattern of broiler chickens under intermittent lighting: Do birds eat in the dark? *Anim. Nutr.* **2018**, *5*, 174–178. [CrossRef]
39. Stahlbaum, C.C.; Rovee-Collier, C.; Fagen, J.W.; Collier, G. Twilight activity and antipredator behavior of young fowl housed in artificial or natural light. *Physiol. Behav.* **1986**, *36*, 751–758. [CrossRef]
40. Xin, H.; Berry, I.; Barton, T.L.; Tabler, G.T. Feeding and Drinking Patterns of Broilers Subjected to Different Feeding and Lighting Programs. *J. Appl. Poult. Res.* **1993**, *2*, 365–372. [CrossRef]
41. Schwean-Lardner, K.; Fancher, B.; Laarveld, B.; Classen, H. Effect of day length on flock behavioural patterns and melatonin rhythms in broilers. *Br. Poult. Sci.* **2014**, *55*, 21–30. [CrossRef]
42. Sherlock, L.; Demmers, T.; Goodship, A.; McCarthy, I.; Wathes, C. The relationship between physical activity and leg health in the broiler chicken. *Br. Poult. Sci.* **2010**, *51*, 22–30. [CrossRef]

animals

MDPI

Article

Developing and Evaluating Poultry Preening Behavior Detectors via Mask Region-Based Convolutional Neural Network

Guoming Li [1], Xue Hui [2], Fei Lin [3] and Yang Zhao [4,*]

1 Department of Agricultural and Biological Engineering, Mississippi State University, Starkville, MS 39762, USA; gl565@msstate.edu
2 College of Energy and Intelligent Engineering, Henan University of Animal Husbandry and Economy, Zhengzhou 450011, China; xh138@msstate.edu
3 Department of Electrical and Computer Engineering, Mississippi State University, Starkville, MS 39762, USA; mojiamenke123@gmail.com
4 Department of Animal Science, The University of Tennessee, Knoxville, TN 37996, USA
* Correspondence: yzhao@utk.edu; Tel.: +1-865-974-6466

Received: 5 August 2020; Accepted: 24 September 2020; Published: 28 September 2020

Simple Summary: Preening is poultry grooming and comfort behavior to keep plumages in good conditions. Automated tools to continuously monitor poultry preening behaviors remain to be developed. We developed and evaluated hen preening behavior detectors using a mask region-based convolutional neural network (mask R-CNN). Thirty Hy-line brown hens kept in an experimental pen were used for the detector development. Different backbone architectures and hyperparameters (e.g., pre-trained weights, image resizers, etc.) were evaluated to determine the optimal ones for detecting hen preening behaviors. A total of 1700 images containing 12,014 preening hens were used for model training, validation and testing. Our results show that the final performance of detecting hen preening was over 80% for precision, recall, specificity, accuracy, F1 score and average precision, indicating decent detection performance. The mean intersection over union (MIOU) was 83.6–88.7%, which shows great potential for segmenting objects of concern. The detectors with different architectures and hyperparameters performed differently for detecting preening birds and thus we need to carefully adjust these parameters to obtain a robust deep learning detector. In summary, deep learning techniques may have a great ability to automatically monitor poultry behaviors and assist welfare-oriented poultry management.

Abstract: There is a lack of precision tools for automated poultry preening monitoring. The objective of this study was to develop poultry preening behavior detectors using mask R-CNN. Thirty 38-week brown hens were kept in an experimental pen. A surveillance system was installed above the pen to record images for developing the behavior detectors. The results show that the mask R-CNN had 87.2 ± 1.0% MIOU, 85.1 ± 2.8% precision, 88.1 ± 3.1% recall, 95.8 ± 1.0% specificity, 94.2 ± 0.6% accuracy, 86.5 ± 1.3% F1 score, 84.3 ± 2.8% average precision and 380.1 ± 13.6 ms·image^{-1} processing speed. The six ResNets (ResNet18-ResNet1000) had disadvantages and advantages in different aspects of detection performance. Training parts of the complex network and transferring some pre-trained weights from the detectors pre-trained in other datasets can save training time but did not compromise detection performance and various datasets can result in different transfer learning efficiencies. Resizing and padding input images to different sizes did not affect detection performance of the detectors. The detectors performed similarly within 100–500 region proposals. Temporal and spatial preening behaviors of individual hens were characterized using the trained detector. In sum, the mask R-CNN preening behavior detector could be a useful tool to automatically identify preening behaviors of individual hens in group settings.

Keywords: poultry; cage-free; preening behavior; mask R-CNN; residual network

1. Introduction

The public and industry have expressed increasing concerns about poultry welfare [1,2]. The performance of natural behaviors is commonly used as a criterion in determination of poultry welfare [3]. Preening is one of natural behaviors of poultry and important for keeping plumages well-groomed in both natural and artificial conditions [4]. During preening, birds use their beaks to distribute lipid-rich oil from the uropygial glands to their feathers, while simultaneously removing and consuming parasites [5,6]. Preening, as a preventive body-surface maintenance behavior, could take a large time budget (~13%) out of the total behavior repertoire of Red Jungle fowl [7], thus being unignorable for welfare evaluation. Proper preening behavior responses help to interpret bird status responding to surroundings. Overall time spent preening and number of preening bouts could reflect environment appropriateness for birds. For example, preening is performed whenever there is nothing more important to do and birds in cages showed more time spent preening than those in nature (26% vs. 15%) [5]. Rearing birds in cages may increase bird boredom and not be suitable for bird welfare. The duration and frequency of preening of individual birds could imply their pleasure/frustration status. Birds having no access to resources (e.g., feeder) may feel frustrated and typically perform short-term and frequent preening [8]. The number of simultaneously preening birds could be an indicator of space sufficiency. If allocated space is not enough for all birds to preen simultaneously, high-ranking birds in social groups have priority to preen first and subordinate ones may need to wait [9]. Spatial distribution of preening birds could help to judge sufficiency of resource allowance as well. For instance, if birds could not access feeders due to insufficient feeder allowance, they would preen near the feeders to displace the mild frustration [4]. These are valuable responses for welfare-oriented poultry production and manually collecting these responses could be time- and labor-consuming. However, there is no available automated tool to extract these preening behavior responses. Precision poultry farming techniques may provide availability to automatically obtain these responses, as various sensors and computer tools have been utilized to detect poultry behaviors [10,11]. Convolutional neural network (CNN) is another potential technology for poultry behavior detection.

Convolutional neural networks have been widely utilized for object detection in agricultural applications [12,13]. With sufficient training, the CNN detectors could precisely detect objects of concern in various environments [14]. Meanwhile, the CNN detectors can be integrated into various vision systems to detect objects non-invasively, which is suitable to detect natural behaviors of poultry without extra interferences. The detection performance of the CNNs is various with architectures. Among them, the mask region-based CNN (mask R-CNN) is an extensive network of faster R-CNN [15]. It was used for detecting pig mounting behaviors [16], apple flowers [13], strawberries [17] and so forth and obtained robust performance on those applications. Besides mask R-CNN, our team also applied single shot detector (SSD), faster R-CNN and region-based fully convolutional network (R-FCN) for detecting floor eggs in cage-free hen housing systems [14]. But from the previous paper [15] and our preliminary test, the mask R-CNN outperformed these network architectures with regard to accuracy because it retained as much object information as possible. Hence, it was selected to detect hen preening behaviors in this case.

Mask R-CNN contains a great number of hyperparameters for training and appropriately tuning/modifying the model is important to develop a robust detector in a customized dataset. The residual network (ResNet) is proposed by He et al. [15] and used as a backbone for the mask R-CNN. Various designs and depths of the ResNets can influence speed and quality for extracting features of input images. Some commonly-used CNN models contain considerable weights that were trained with some benchmark datasets, such as common objects in context 'COCO' [18] and ImageNet [19]. The weights pre-trained with COCO and ImageNet dataset were hereafter named as pre-trained COCO weights and pre-trained ImageNet weights. To apply the models into customized datasets, one efficient solution is to transfer the pre-trained weights learned previously into parts of the model and only trained the rest parts. Such transfer learning could save training time and simultaneously not compromise network performance [20]. Before developing deep learning models, image resizers are typically used to uniform sizes of input images in benchmark datasets, in which sizes of images are various due to different photographing conditions. Inappropriate resizing strategies may downgrade detection performance. For example, resizing large images into small ones may increase processing speed but risk missing small objects in the resized images [21]; and enlarging small images into large ones with changed length-to-width ratios could distorted shapes and features of objects of concern [22]. Insufficient region proposals may lead to missing target objects while excessive proposals may downgrade processing speed [23]. However, it is uncertain which backbone architecture is better for detecting preening birds and which hyperparameters are more efficient to develop the detectors.

The objective of this research was to develop mask R-CNN preening behavior detectors using brown hens as examples. The brown hens lay brown-shell eggs accounting for a large share (>90% in Europe and >70% in China) of the global egg market [24]. The backbone architecture and hyperparameters, including pre-trained weight, image resizer and regions of interest (ROI), were modified to construct an optimal detector for the detection purpose. The backbone architectures of residual networks (ResNet) were ResNet18, ResNet34, ResNet50, ResNet101, ResNet152 and ResNet1000. The trainings included without pre-trained weights, with the pre-trained COCO weights and with the pre-trained ImageNet weights. The modes of image resizers were 'None,' 'Square' and 'Pad64'. Numbers of ROIs were 30, 100, 200, 300, 400 and 500. With the trained detector, hen preening behaviors were quantified as well.

2. Materials and Methods

2.1. Housing, Animals and Management

The experiment was conducted at the U.S. Department of Agriculture (USDA) Poultry Research Unit at Mississippi State, USA and all procedures in this experiment were approved by the USDA-ARS Institutional Animal Care and Use Committee at Mississippi State, USA. Thirty Hy-line Brown hens at 38 weeks of age were placed in a pen, measuring 2.5 m long × 2.2 m wide. Nest boxes, feeders and drinkers were equipped in the pen. Fresh litter was spread on the floor before bird arrival. Commercial feed was provided ad libitum. Temperature, light program and light intensity were, respectively, set to 24 °C, 16L:8D (light ON at 6:00 am and OFF at 10:00 pm) and 20 lux at bird head level.

2.2. Data Acquisition

A night-vision network camera (PRO-1080MSB, Swann Communications U.S.A Inc., Santa Fe Springs, LA, USA) was mounted in the middle of the pen and at ~2 m above the ground to capture top-view videos. Hen activity was continuously monitored and videos were stored in a digital video recorder (DVR-4580, Swann Communications U.S.A Inc., Santa Fe Springs, LA, USA). The video files were recorded with a resolution of 1280 × 720 pixels at a sample rate of 25 frames per second (fps) and converted to image files (.jpg) using Free Video to JPG Converter (ver. 5.0).

2.3. Preening Behavior Definition and Labelling

The definition of preening was that a bird grooms its feathers on different body parts, including breast, throat, belly, shoulder, wing, back, tail and vent [6,25–27]. Based on the definition, we manually labeled each preening hen that had the features in Figure 1. It should be noted that this study examined preening behavior with beak only and the preening behavior with foot [26] was not considered. A total of 48 h of videos, 16 h in one day, was used. Images with at least 1-min intervals were selected [12] and the images containing preening hens were used for the labelling, resulting in totally 1700 images from three-day videos. The labelling was conducted in an open-source labeling software (VGG Image Annotator, VIA 2.0.4). A protocol of labeling preening birds that had the features in the preening definition was set. The dataset was split into two parts and two experienced labelers labeled respective parts of images following the protocol. Then they mutually checked the labeled results to ensure that the labels were correct.

Figure 1. Sample pictures of preening hens. The preening birds were manually cropped from original images.

2.4. Network Description

The mask R-CNN consists of a backbone to extract features from an input image, a region proposal network (RPN) to propose ROI and a detection head for object detection and instance segmentation (Figure 2).

Each input image is first resized into a proper size using an image resizer. A ResNet and a feature pyramid network (FPN) are used to construct the backbone to extract features from the resized image. The ResNet is a bottom-up convolution network and divided into five stages of convolutions (C1–C5) [15]. With higher stages of convolution, the sizes of resultant maps become smaller and higher-level semantics are retained. The FPN is a top-down convolution network and generates five scales of feature maps (P2–P6), which are resulted from the C2–C5 maps, respectively. The C2–C5 and P2–P6 maps are laterally connected with a convolution of $1 \times 1 \times 256$ and up-sampling with the size of (2, 2). The P6 map is processed from the P5 map with a max pooling of [(1, 1), 2]. The ResNet-FPN structure facilitates the extraction of both lower- and higher-level semantics, which are critical for instance segmentation with regards to objects having various scales in an image. The P2–P5 maps are concatenated to form feature maps for detection head, while the P2–P6 maps are combined and go through a convolution of $3 \times 3 \times 256$ to process a map for RPN.

In the RPN, an anchor generator generates the anchors with 5 scales of 32, 64, 128, 256 and 512 and 3 ratios of 0.5, 1 and 2. These anchors are tiled onto the map generated from the P2–P6 maps and then a series of candidate boxes synthesized with objectness and bounding box deltas are proposed. With the non-maximum suppression (NMS) rule, unnecessary boxes are filtered out and ROIs are retained. The ROIs are finally projected onto the feature maps to position objects of interest using ROI Align operation. The ROI Align uses the bilinear function to maintain float coordinates and makes pixel-wise prediction more accurate than the ROI Pooling in the faster R-CNN, in which float coordinates are typically quantized and valuable pixel information may lose.

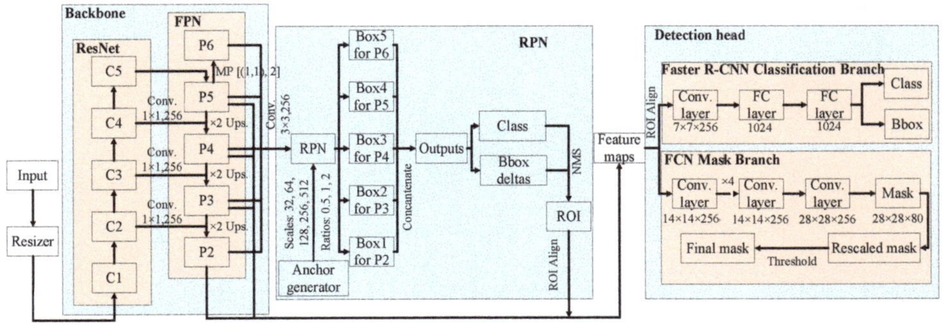

Figure 2. Network structure of mask region-based convolutional neural network (mask R-CNN). ResNet is a residual network; FPN is a feature pyramid network; RPN is a region proposal network; ROI is region of interest; NMS is non-maximum suppression; FC layer is fully-connected layer; Bbox is bounding box; FCN is fully-connected network; C1–C5 are convolutional stages 1 to 5 in the ResNet; P2–P6 are feature maps in the FPN; Box1–Box5 are proposed boxes with various scales and ratios after the RPN; Conv. $1 \times 1{,}256$ is the convolution with the kernel size of (1, 1) and depth of 256; MP [(1, 1), 2] is max pooling with the size of (1, 1) and stride of 2; ×2 Ups. is upsampling with the size of (2, 2); Conv. $3 \times 3 \times 256$ is the convolution with the kernel size of (3, 3) and depth of 256; $7 \times 7 \times 256$ is the size (length of 7, width of 7 and depth of 256) of convolution layers; 1024 is the number of neurons in the FC layer; $14 \times 14 \times 256$ is the size (length of 14, width of 14 and depth of 256) of convolution layers; ×4 is the repeated operations of the previous layer for 4 times; $28 \times 28 \times 256$ is the size (length of 28, width of 28 and depth of 256) of convolution layers; $28 \times 28 \times 80$ are 80 target masks with the size of 28 in length and 28 in width.

The detection head comprises three branches that are object classification branch, bounding box regression branch and object instance segmentation branch. The first two branches belong to the faster R-CNN classification branch and the third branch is the fully-connected network (FCN) mask branch. Various sizes of feature patches are proposed after the above-mentioned procedures and resized to consistent sizes using another ROI Align operation, which can again retain more pixels than the ROI Pooling. For the faster R-CNN branch, the resized feature patches go through a convolution layer of $7 \times 7 \times 256$ and two 1024-neuron fully-connected (FC) layers to predict object scores and refine object locations. As for the FCN branch, the patches undergo several convolution layers of $14 \times 14 \times 256$ and a de-convolution layer of $28 \times 28 \times 256$. Eighty 28×28 candidate masks are processed and rescaled according to the image size. Each pixel with the score being greater than 0.5 is assigned to the object of concern to generate the final binary mask, which is visualized together with the bounding box and class name.

2.5. General Workflow of Detector Training, Validation and Testing

Figure 3 shows the overall process of training, validation and testing. Training data was input into the mask R-CNN detectors for training and the training loss was continuously calculated during the training process. The training detectors were stored in specific training iteration periodically and validated with the validation set. The training and validation losses were compared. If training and validation losses kept decreasing, it meant that the detectors were underfitted and needed more training. If the training loss decreased while the validation loss increased, it meant that the detectors were overfitted and the training process needed to be stopped [28]. With the final saved detectors, the hold-out testing data was used to evaluate the detector performance on preening detection. The computing system used for detector training, validation and testing computing was equipped with 32 GB RAM, Intel(R) Core (TM) i7-8700K processor and NVIDIA GeForce GTX 1080 GPU card (Dell Inc., Round Rock, TX, USA).

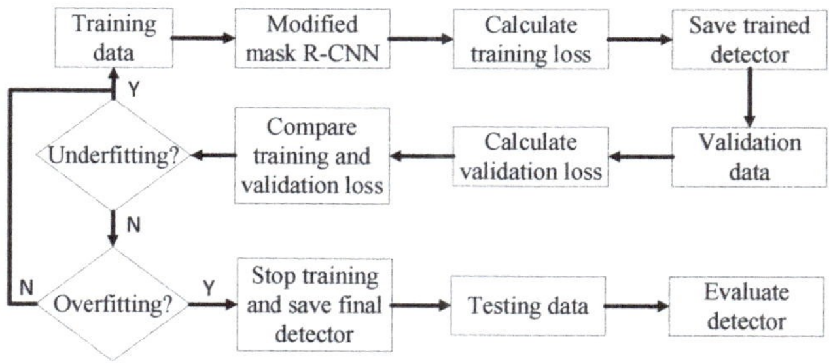

Figure 3. Illustration of the training, validation, and testing process. Mask R-CNN is mask region-based convolutional neural network. "Y" means that judgement is true and "N" means that judgement is false.

Table 1 shows the data distribution for training, validation and testing. Labeled images were described in Section 2.3 with 1175 images for training, 102 for validation and 423 for testing, resulting in 8464 labeled hens for training, 762 for validation and 2788 for testing. The training, validation and testing data came from three different days, respectively and those images had at least 1-min intervals. Therefore, they were thought to have sufficient variations for detector development.

Table 1. Data distribution for training, validation and testing.

Items	Training	Validation	Testing
Hen age (day)	266	267	268
Images	1175	102	423
Number of preening hens	8464	762	2788
Number of non-preening hens	26,786	2298	9902

The losses for training and validation included total loss, detection head class loss, detection head bounding box loss, detection mask loss, RPN bounding box loss and RPN class loss. The six types of losses were reported by He et al. [15] and reflected how much deviation there was between prediction and ground truth (Figure 4). Except for total loss, the other five types of losses corresponded to the three outputs in the detection head and two outputs in the RPN and the total loss was the sum of the five losses. A smaller loss indicated a better prediction. For instance, as loss samples shown in Figure 4, the training losses kept decreasing, while most of the validation losses decreased before 9×10^3 iterations and had a rebound increase after 9×10^3 iterations. Therefore, the training process was stopped at the 9×10^3th iteration to avoid overfitting and the detectors were saved accordingly.

2.6. Modifications for Detector Development

The modifications for the detector development involved ResNet architecture, pre-trained weight, image resizer and number of ROI. The detectors with the following modifications was trained as described in Section 2.5 and the modification with optimal testing performance was used to develop the preening behavior detectors. The following modifications were trained with the default settings of mask R-CNN that were ResNet101, pre-trained COCO weights, 'Square' image resizer mode and 200 ROIs, unless specified in the sections. As for other hyperparameters for training, we followed the default settings recommended by Abdulla [22].

2.6.1. Residual Network Architecture

The ResNet was proposed by He, et al. [29]. Sufficiently extracting semantics in the C2–C5 stages was critical for detection performance. Six ResNet architectures that were ResNet18, ResNet34, ResNet50, ResNet101, ResNet152 and ResNet1000 were embedded into the mask R-CNN backbone for training (Table 2). The number beside 'ResNet' indicates the number of layers in the architecture. The ResNets with less than 50 layers were constructed with normal blocks (Figure 5a), while those with more than 50 layers were stacked with bottleneck blocks (Figure 5b), which can reduce computational complexity with increasing layers in the ResNet. The original mask R-CNN was built with ResNet50 or ResNet101.

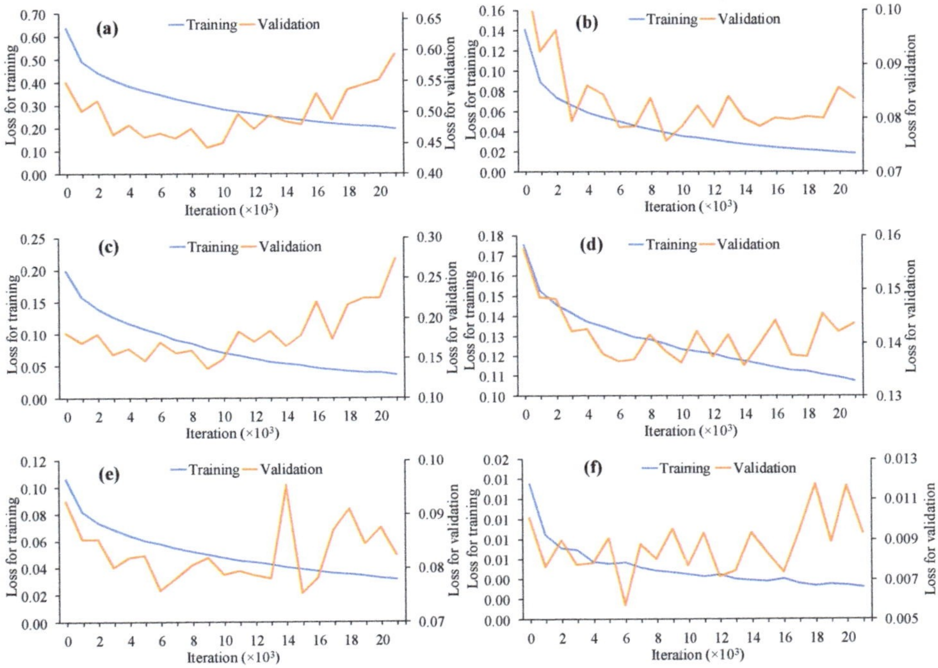

Figure 4. Samples of training and validation losses during training process. (**a**) Total loss; (**b**) detection head bounding box loss; (**c**) detection head class loss; (**d**) detection head mask loss; (**e**) region proposal network bounding box loss; and (**f**) region proposal network class loss.

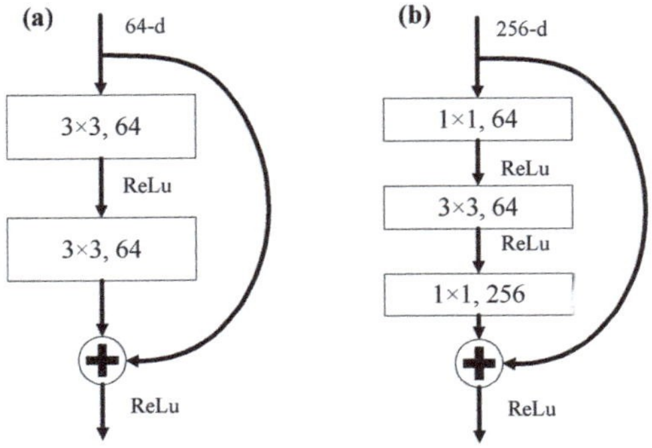

Figure 5. Examples of building block with shortcut connections for the residual network (ResNet). (**a**) Normal block at the C2 stage of the ResNet18 and ResNet34; and (**b**) bottleneck block at the C2 stage of the ResNet50-ResNet1000. ReLu is rectified linear units; 64-d is depth of 64; and 256-d is depth of 256. The figure was redrawn from He, et al. [29].

Table 2. Detailed architectures for residual network (ResNet).

Stage of Convolution	ResNet18	ResNet34	ResNet50	ResNet101	ResNet152	ResNet1000
C1	7 × 7, 64, stride 2 3 × 3 max pooling, stride 2					
C2	$\begin{bmatrix} 3\times3, 64 \\ 3\times3, 64 \end{bmatrix} \times 2$	$\begin{bmatrix} 3\times3, 64 \\ 3\times3, 64 \end{bmatrix} \times 3$	$\begin{bmatrix} 1\times1, 64 \\ 3\times3, 64 \\ 1\times1, 256 \end{bmatrix} \times 3$	$\begin{bmatrix} 1\times1, 64 \\ 3\times3, 64 \\ 1\times1, 256 \end{bmatrix} \times 3$	$\begin{bmatrix} 1\times1, 64 \\ 3\times3, 64 \\ 1\times1, 256 \end{bmatrix} \times 3$	$\begin{bmatrix} 1\times1, 64 \\ 3\times3, 64 \\ 1\times1, 256 \end{bmatrix} \times 248$
C3	$\begin{bmatrix} 3\times3, 128 \\ 3\times3, 128 \end{bmatrix} \times 2$	$\begin{bmatrix} 3\times3, 128 \\ 3\times3, 128 \end{bmatrix} \times 4$	$\begin{bmatrix} 1\times1, 128 \\ 3\times3, 128 \\ 1\times1, 512 \end{bmatrix} \times 4$	$\begin{bmatrix} 1\times1, 128 \\ 3\times3, 128 \\ 1\times1, 512 \end{bmatrix} \times 4$	$\begin{bmatrix} 1\times1, 128 \\ 3\times3, 128 \\ 1\times1, 512 \end{bmatrix} \times 22$	$\begin{bmatrix} 1\times1, 128 \\ 3\times3, 128 \\ 1\times1, 512 \end{bmatrix} \times 248$
C4	$\begin{bmatrix} 3\times3, 256 \\ 3\times3, 256 \end{bmatrix} \times 2$	$\begin{bmatrix} 3\times3, 256 \\ 3\times3, 256 \end{bmatrix} \times 6$	$\begin{bmatrix} 1\times1, 256 \\ 3\times3, 256 \\ 1\times1, 1024 \end{bmatrix} \times 6$	$\begin{bmatrix} 1\times1, 256 \\ 3\times3, 256 \\ 1\times1, 1024 \end{bmatrix} \times 23$	$\begin{bmatrix} 1\times1, 256 \\ 3\times3, 256 \\ 1\times1, 1024 \end{bmatrix} \times 22$	$\begin{bmatrix} 1\times1, 256 \\ 3\times3, 256 \\ 1\times1, 1024 \end{bmatrix} \times 247$
C5	$\begin{bmatrix} 3\times3, 512 \\ 3\times3, 512 \end{bmatrix} \times 2$	$\begin{bmatrix} 3\times3, 512 \\ 3\times3, 512 \end{bmatrix} \times 3$	$\begin{bmatrix} 1\times1, 512 \\ 3\times3, 512 \\ 1\times1, 2048 \end{bmatrix} \times 3$			$\begin{bmatrix} 1\times1, 512 \\ 3\times3, 512 \\ 1\times1, 2048 \end{bmatrix} \times 247$
	Average pooling, 1000-d FC, softmax					

Note: ResNet18-ResNet1000 are residual network with 18–1000 layers of convolution; C1–C5 are convolutional stages 1 to 5 in the ResNet; and FC is fully-connected.

2.6.2. Pre-Trained Weight

The mask R-CNN was pre-trained with the benchmark datasets of COCO [18] and ImageNet [19] and obtained pre-trained weights. The trainings included without pre-trained weights, with pre-trained COCO weights and with pre-trained ImageNet weights. The training with the pre-trained weights only involved the heads of FPN, RPN and detection branches, which contained 28 items, while the full layer training without pre-trained weights was related to every layer in the detectors, which contained 236 items in total.

2.6.3. Image Resizer

To obtain uniform size of images for detector development, we need to resize the input images to the same size. Appropriate image resizers could improve processing speed and retain as much pixel-wise information as possible [22]. Three modes of resizers were compared, which were 'None,' 'Square' and 'Pad64'. In the 'None' mode, input images (1280 × 720 pixels) were neither resized nor padded. In the 'Square' mode, input images were resized from 1280 × 720 pixels to 1024 × 1024 pixels and zeros were used to pad blank areas of resized images. In the 'Pad64' mode, input images were resized from 1280 × 720 pixels to 1280 × 768 pixels and the differences were padded with zeros. Resized sample images with the three modes of resizing are shown in Figure 6.

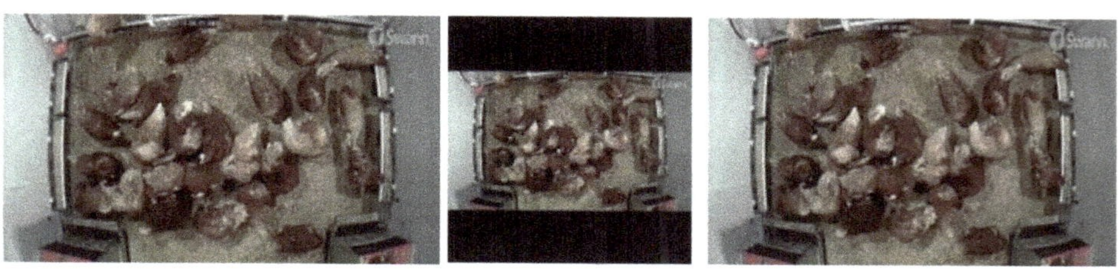

'None' mode: 1280 × 720 pixels | 'Square' mode: 1024 × 1024 pixels | 'Pad64' mode: 1280 × 768 pixels

Figure 6. Resized sample images with three modes of image resizers.

2.6.4. Proposed Regions of Interest

Target preening birds may be ruled out in a feature map with insufficient ROIs, resulting in miss-identification of preening birds, while processing speed may decrease using a map with excessive

ROIs [23]. The detectors were trained with 30, 100, 200, 300, 400 and 500 ROIs and the performance was compared.

2.7. Evaluation Metrics

After the detectors were trained and validated, the hold-out testing set was used for evaluating the trained detectors as described in Section 2.5. To determine whether a preening hen had been correctly segmented, the intersection over union (*IOU*) for each predicted hen was computed using overlap and union pixels of the ground truth and prediction (Equation (1)). An *IOU* greater than 0.5 in this case means the detectors segmented and detected a preening hen correctly.

$$IOU[\%] = \frac{(pixels \in ground\ truth) \cap (pixels \in prediction)}{(pixels \in ground\ truth) \cup (pixels \in prediction)} \times 100\% \quad (1)$$

The mean *IOU* (MIOU) was used to evaluate overall segmentation performance of the detectors and calculated in Equation (2).

$$MIOU = \frac{\sum_{i=1}^{n} IOU_i}{n} \quad (2)$$

where IOU_i is the *IOU* for the *i*th preening hen and *n* is the total number of preening hens.

Precision, recall, specificity, accuracy and F1 score for detecting each preening hen in the images were calculated using Equations (3)–(7). Precision is the percentage of true preening cases in all detected preening cases. Recall is the percentage of the true preening cases in all manually-labeled preening cases. Specificity is the percentage of true non-preening cases in all manually-labeled non-preening cases. Accuracy is the percentage of true preening and non-preening cases in all cases. F1 score is the harmonic mean of precision and recall and a balance metric on comprehensively evaluating false preening and non-preening cases. For all five metrics, a closer to 100% value reflects a better performance of the detectors.

$$Precision[\%] = \frac{TP}{TP + FP} \times 100\% \quad (3)$$

$$Recall[\%] = \frac{TP}{TP + FN} \times 100\% \quad (4)$$

$$Specificity[\%] = \frac{TN}{TN + FP} \times 100\% \quad (5)$$

$$Accuracy[\%] = \frac{TN + TP}{TN + TP + FN + FP} \times 100\% \quad (6)$$

$$\text{F1 score } [\%] = 2 \times \frac{Precision \times Recall}{Precision + Recall} \times 100\% \quad (7)$$

where *TP* is true positive, that is, number of cases that a detector successfully detects existent preening hens in an image with *IOU* greater than 0.5; *FP* is false positive, that is, number of cases that a detector reports non-existent preening hens in an image or *IOU* is less than 0.5; *FN* is false negative, that is, number of cases that a detector fails to detect existent preening hens in an image; and *TN* is true negative, that is, number of cases that non-preening hens are reported by both a detector and manual label.

Average precision (*AP*) summarizes the shape of the precision-recall curve and is defined as the mean precision at a set of 11 equally-spaced recall levels [0, 0.1, . . . , 1] [30]. The precision-recall curve is produced according to the predicted confidence level. Increasing the confidence may reduce false positives but increase false negative, resulting in increasing precision and decreasing recall. A closer to 100% *AP* indicates a more generalized detector to detect objects with various confidence. The calculation of the *AP* is shown in Equation (8).

$$AP[\%] = \frac{1}{11} \sum_{r \in \{0, 0.1, \ldots, 1\}} P_{interp}(r) \quad (8)$$

where r is level of recall at $\{0, 0.1, \ldots, 1\}$; and $P_{interp}(r)$ is the interpolated precision in the precision-recall curve when recall is r.

The interpolated precision is the maximum value within one piece of a wiggle-shape curve (Equation (9)).

$$P_{interp}(r) = \max_{\tilde{r}:\tilde{r}\geq r} P(\tilde{r}) \tag{9}$$

where $\tilde{r}$ is the recall within a wiggle piece; and $P(\tilde{r})$ is the measured precision at recall $\tilde{r}$.

The processing time reported by Python 3.6 was used to evaluate the processing speed of the detectors for processing 423 images. The processing speed (ms·image^{-1}) was obtained by dividing the total processing time with 423 images.

2.8. Sample Detection

We finally deployed the detector trained with ResNet101, pre-trained COCO weights, 'Square' mode and 200 ROIs, after the performance comparison. We continuously detected hen preening behaviors for half hour in week 38 of bird age. A segmented image based on traditional Otsu's thresholding [31] was used to compare the result of preening instance segmentation using the mask R-CNN detector. The hen preening behaviors at 6:00 am–6:30 am were characterized as time spent preening (min·hen^{-1}), number of preening bouts (bouts·hen^{-1}), average preening duration (min·bout^{-1}), frequency of preening duration, number of birds simultaneously preening and spatial distribution of preening birds. Spatial location of preening birds was plotted in a heat map. To construct a heat map, a mesh grid was firstly constructed onto the pen map based on the dimension of the pen, in which the gird size was set to 10 pixels. Then a Standard Gaussian Kernel Density Estimation Function was run onto the center of each grid in the map and the preening frequency in each grid was calculated by Equation (10). Finally, the density map was visualized using Matplotlib, an open-source visualization library. The cooler-color (i.e., dark blue) areas in the map represented the areas where birds performed preening more often, while the warmer-color (i.e., dark red) areas were the areas where birds were less likely to preen.

$$P = \sum_{i=1}^{n} \frac{1}{\sqrt{2\pi}} e^{-d_i^2/2} \tag{10}$$

where P is the probability in Standard Gaussian Distribution curve; n is the total number of grids in the entire image; and d_i is the pixel-representing distance between the grid center and ith detected preening bird center.

3. Results

3.1. Sample Detection

A sample detection is shown in Figure 7. Individual preening birds could be detected and segmented separately using the mask R-CNN detector, while some segmented hens by traditional thresholding method were cohesive and mixed with the background due to similar features with the background. Therefore, we applied the CNN detector to detect preening hens in this case. Meanwhile, as the deep learning outperformed the traditional image processing with regards to object segmentation, it may be a better choice for some research/application purposes (e.g., bird activity analysis).

(a) (b) (c)

Figure 7. Sample detection. (**a**) Original image; (**b**) preening instance segmentation using the mask region-based convolution neural network detector; and (**c**) image segmentation using Otsu's thresholding. Some preening birds in the Figure 7b are masked with different colors.

3.2. Performance of Various Residual Networks

Table 3 shows the performance of various ResNets on preening detection. The six ResNets had similar segmentation performance of preening birds as indicated by similar MIOU (87.3–87.8%). The ResNet18 had middle performance but the second fastest processing speed (364.8 ms·image^{-1}). The ResNet34 had the highest precision (88.5%) but the second lowest recall (86.2%) and AP (83.1%). The ResNet50 had the lowest recall (85.3%), accuracy (93.5%), F1 score (84.9%), AP (81.4%) and processing speed (342.9 ms·image^{-1}). The ResNet101 had the highest accuracy (95.0%) and F1 score (88.1%). The ResNet152 had the lowest precision (83.1%) but the highest AP (85.7%). The ResNet1000 had the third lowest precision (84.5%) and slowest processing speed (393.2 ms·image^{-1}). Overall, the six ResNets had strengths and weaknesses in different detection performance. Since the ResNet101 is a popular network in other agriculture applications, it was selected to develop the detector.

Table 3. Performance of various residual networks on preening detection.

ResNet	MIOU (%)	Precision (%)	Recall (%)	Specificity (%)	Accuracy (%)	F1 Score (%)	AP (%)	Processing Speed (ms·image^{-1})
ResNet18	87.4	87.2	87.1	96.7	94.7	87.1	83.6	364.8
ResNet34	87.4	88.5	86.2	97.0	94.8	87.3	83.1	378.4
ResNet50	87.8	84.4	85.3	95.7	93.5	84.9	81.4	342.9
ResNet101	87.4	87.7	88.4	96.7	95.0	88.1	83.5	386.0
ResNet152	87.4	83.1	90.1	95.1	94.1	86.5	85.7	387.7
ResNet1000	87.4	84.5	90.8	95.6	94.6	87.6	85.6	393.2

Note: ResNet is residual network; ResNet18-ResNet1000 is the ResNet with 18–1000 layers; MIOU is mean intersection over union; and AP is average precision.

3.3. Performance of the Detectors Trained with Various Pre-Trained Weights

Table 4 shows the detection performance of the mask R-CNN detectors trained with various pre-trained weights. The detector trained with pre-trained ImageNet weights had the low performance of preening detection and segmentation among the three trainings, except for the specificity and precision. The detectors trained without pre-trained weights and with pre-trained COCO weights had similar detection performance, while the former training took ~50% more time and ~70% more computer memory. Therefore, the detector was trained with the pre-trained COCO weights in this case.

Table 4. Performance of the detectors trained with various pre-trained weights.

Training	MIOU (%)	Precision (%)	Recall (%)	Specificity (%)	Accuracy (%)	F1 Score (%)	AP (%)	Processing Speed (ms·image^{-1})
w/o pre-trained weights	88.7	80.3	92.3	93.9	93.6	85.9	87.5	379.0
w/pre-trained COCO weights	87.2	83.4	91.3	94.5	93.8	87.2	86.7	382.9
w/pre-trained ImageNet weights	83.6	81.2	83.1	94.9	92.5	82.2	80.0	413.7

Note: COCO is common object in context; MIOU is mean intersection over union; and AP is average precision. 'w/o' and 'w/' indicate 'without' and 'with', respectively.

3.4. Performance of Various Image Resizers

Table 5 shows the performance of various image resizers. Similar MIOU and accuracy were observed for the three modes of resizers. The 'Square' mode had the lowest recall (86.3%) and F1 score (86.6%) but the highest specificity (96.6%). The 'Pad64' had the lowest precision (84.2%) and specificity (95.6%) but the highest recall (90.1%) and processing speed (383.3 ms·image^{-1}). It should be noted that except for processing speed and recall, performance differences among the resizers were mostly less than 3%. As there was no obvious strength of detection performance among the resizers, the default resizers ('Square' mode) was deployed to develop the final detector.

Table 5. Performance of various image resizers.

Mode of Image Resizer	MIOU (%)	Precision (%)	Recall (%)	Specificity (%)	Accuracy (%)	F1 Score (%)	AP (%)	Processing Speed (ms·image^{-1})
None	87.2	85.3	88.4	96.0	94.4	86.8	84.6	377.7
Square	87.6	86.9	86.3	96.6	94.5	86.6	86.7	377.8
Pad64	87.0	84.2	90.1	95.6	94.5	87.0	86.3	383.3

Note: MIOU is mean intersection over union; and AP is average precision.

3.5. Performance of the Detectors Trained with Various Numbers of Regions of Interest

The MIOU, accuracy and F1 score were similar among different numbers of ROIs (Table 6). The detector trained with 30 ROIs had the lowest recall (79.3%) and AP (75.8%) but the highest precision (92.5%) and specificity (98.2%). With more than 30 ROIs, the precision, recall, specificity and AP, respectively, ranged from 82.8–85.8%, 87.2–90.7%, 95.0–96.3% and 84.9–86.5%. The processing speed (378.2–390.7 ms·image^{-1}) did not absolutely increase as more ROIs were used. Because there was no absolute improvement of detection performance with increasing ROIs (>30), the default ROIs of 200 was used in the final training.

Table 6. Performance of the detectors trained with various numbers of regions of interest (ROI).

Number of ROI	MIOU (%)	Precision (%)	Recall (%)	Specificity (%)	Accuracy (%)	F1 Score (%)	AP (%)	Processing Speed (ms·image^{-1})
30	87.5	92.5	79.3	98.2	94.2	85.4	75.8	378.2
100	87.1	84.2	89.1	95.5	94.2	86.6	84.9	379.4
200	86.9	85.8	89.5	96.0	94.6	87.6	85.4	378.1
300	87.2	85.7	87.2	96.3	94.4	86.4	83.8	390.7
400	87.6	82.7	90.3	95.0	94.0	86.3	86.5	382.0
500	87.0	82.8	90.7	95.0	94.1	86.6	86.2	378.8

Note: MIOU is mean intersection over union; and AP is average precision.

3.6. Preening Behavior Measurement via the Trained Detector

Figure 8 shows the preening behavior measurement in half hour via the trained detector. The hen spent on average 18.1 min, 106 bouts and 0.23 min·bout^{-1} on preening. For over 90% of the time, the hens preened for less than 30 sec. A hen preened for up to 20.5 min within a preening event. Ten birds choosing to simultaneously preen took up the most proportion (16.9%) and the overall frequency was distributed in a shape of normal distribution. The hens spent more time preening at the top left corner of the pen. The hotspot was caused by multiple preening birds. Some birds may finish the preening and leave for eating/drinking while others may enter that area for preening.

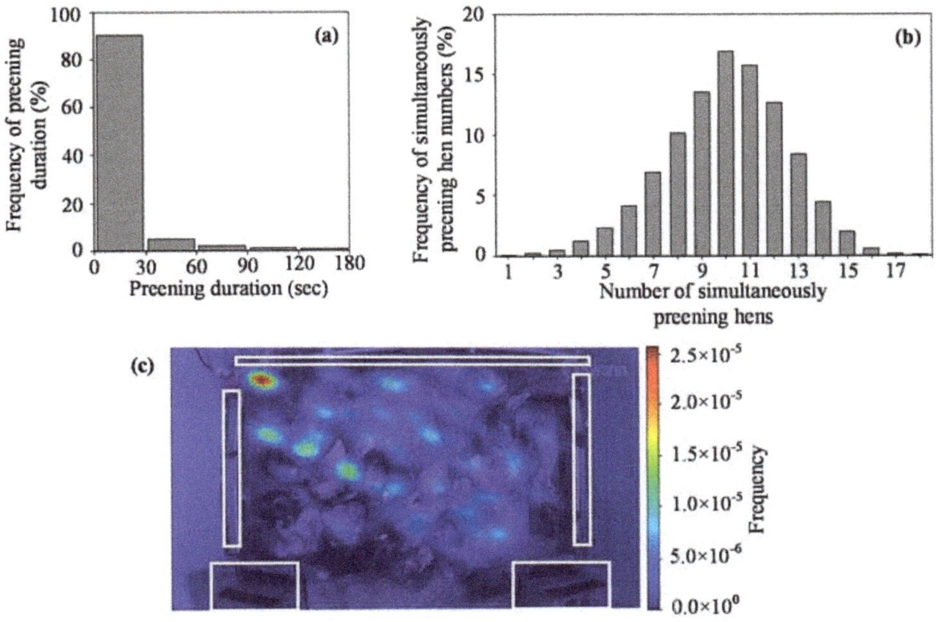

Figure 8. Preening behavior measurement during 6:00–6:30 am for 38-week-old hens. (**a**) Frequency of preening duration. (**b**) Frequency of simultaneously preening numbers. (**c**) Heat map for the location of preening birds. The drinker, feeder and nest box are marked as the white rectangles on the top, middle and bottom, respectively. The non-unit frequency represents the probability for birds preening at specific locations and is calculated by Standard Gaussian Kernel Density Estimation Function.

4. Discussion

4.1. Ambiguous Preening Behavior

Hens can preen various parts of their body. In the images from the single camera and camera angle, some birds were hard to manually tell whether they were preening (Figure 9), which were ~3% of all hens in images. Those questionable hens were not labeled in this case and only the hens with clear preening features as mentioned in Figure 1 were labeled. Although this could make the detectors accurately detect preening hens with obvious features, the detectors still inevitably became ambiguous on detecting those questionable hens, especially the birds preening their chest or pecking ground. That could compromise detection performance. To reduce the confusion and improve detection performance, multiple cameras with multiple angles may be considered to capture different views of preening birds. To exemplify the applications of mask R-CNN on poultry preening behavior detection, we trained the detectors with images only containing brown hens. However, the detector may also be trained with other images containing other types of chickens, which can extend application range. Based on our previous experiment, deep learning networks could be generalized to different light intensities, backgrounds, object colors, object numbers and object sizes, as long as they were fed and trained with enough sample images [14].

Figure 9. Sample pictures of questionable birds. The birds were manually cropped from original images. They were not labeled as preening birds and used for model development.

As shown in Figure 8a, birds typically preened for less than 30 s and some birds even did it shorter (<1 s). With that regard, if the accuracy is acceptable, higher processing speed is still preferred since it may cover more prompt preening behaviors.

4.2. Segmentation Method Comparison

The compared segmentation method, despite not being the most state-of-the-art, is still commonly used for bird activity evaluation [11,32]. Pairs of adjacent images were compared to get the difference images and then the resulting images were binarized using the image processing methods. Activity index may reflect status of birds responding to surroundings and is important for poultry production. This parameter was extracted with traditional image segmentation methods. However, based on our current test, the traditional methods may result in cohesive maps. Although delicate adjustment may help to solve the issue, it requires intensive labor and has poor generalization. The mask R-CNN could solved the problem and may facilitate the bird activity evaluation, thus being recommended in this case.

4.3. Architecture Selection

The mask R-CNN is not the most state-of-the-art architecture. However, as agricultural engineers, our major goal is not to pursue the most advanced technique regardless but to seek reliable solutions to facilitate agricultural productions. With that regard, mask R-CNN was widely used in different areas and commercialized to accurately detect objects of concern, which may be acceptable for farmers, thus being our solution to detect poultry preening. Meanwhile, based on our preliminary test, the mask R-CNN outperformed its counterparts (e.g., SSD, faster R-CNN and R-FCN) in terms of accuracy, because the mask R-CNN retained as many pixels as possible using FPN and ROI align. Taking these into consideration, we decided to test the mask R-CNN systematically and seek the optimal modifications of the mask R-CNN.

4.4. Performance of Various Residual Network

Various ResNet architectures had different performance on preening detection. The efficiency of instance segmentation mainly relied on the mask detection architecture (FCN in this case) in the mask R-CNN detector [15,33] and the same mask detection architecture among various ResNets caused similar MIOU. Although the ResNet101 had slightly better or similar precision, recall, specificity, accuracy, F1 score and AP, the differences of the performance were small (2–4%). The ResNet-FPN backbone was proposed to extract ROI features from different levels according to their scales within an images and ResNet with more layers may theoretically improve the extraction and further detection performance [29,34]. The scales in some common images were partitioned at three levels that were <1024 pixels, 1024–9216 pixels and >9216 pixels corresponding to small, medium and large objects, respectively [35]. Compared with those, the scales of preening birds ranged from 7765 to 17,353 pixels, in which object areas in an image were relatively consistent. That may be the reason for why we cannot get significant improvement with ResNets having more layers.

Appropriate design of the CNN architecture could improve processing speed as more layers are stacked onto the architecture. The processing speed increased as the ResNet layers increased from 18 to 34 and 50 to 1000. However, it decreased as the ResNet increased from 34 layers to 50 layers. As shown in Figure 4, the ResNet18–ResNet34 were built with normal blocks, while the ResNet50–ResNet1000 were constructed with bottleneck blocks. The latter design can reduce the computational complexity and further increase the processing speed in deeper ResNets (≥50 layers) [29]. In sum, proper CNN architectures are critical for developing a robust and efficient detector as it can affect detection performance.

4.5. Performance of the Detectors Trained with Various Pre-Trained Weights

Modern CNNs are massive architectures containing considerable parameters to be trained, thus, efficiently training the inner structure is critical. Transfer learning could be a solution. Compared with the performance between the trainings without pre-trained weights and with pre-trained COCO weights, transfer learning could save training time without compromising detection performance. The latter training only involved the heads of FPN, RPN and detection branches, containing high-level semantics [36]. Such semantics may be more important for instance segmentation and object detection than low-level generic features extracted by the bottom architecture of the detectors [15]. That is why the detectors with the two trainings showed similar performance. As for transfer learning, various pre-trained weights among benchmark datasets can result in various performance. Perhaps, the pre-trained COCO weights had more similarity for preening hens than the pre-trained ImageNet weights, resulting in better efficiency of transfer learning and better performance for the former pre-trained weights [18–20].

4.6. Performance of Various Image Resizers

The resizers in this case had similar detection performance. The original image size was 1280 × 720 pixels and the size after resizing was 1024 × 1024 pixels for the 'Square' mode and 1280 × 768 pixels for the 'Pad64' mode. Most of the sizes were the multipliers of 64, which can ensure smooth scaling of feature maps up and down at the six levels of the FPN and reduce information loss [22]. That could result in the similar performance. Furthermore, the shapes of preening birds were not distorted before and after resizing, which made the detectors learn consistent features of preening birds and generate similar results. Reducing input image sizes indeed can help to cut processing time [23].

4.7. Performance of the Detectors Trained with Various Numbers of Regions of Interest

Detection performance varied with the proposed numbers of ROI. When less ROIs (<30) were proposed, some candidate preening hens may be ruled out, resulting in low recall and AP [23]. Meanwhile, fewer non-preening birds may be wrongly recognized as preening birds with less ROIs, causing higher precision and specificity. However, these trends disappeared when the ROIs were more than 100. Perhaps, more than 100 ROIs were sufficient to cover possible preening hens for the detection in this case. Processing speed did not absolutely increase with advanced ROIs, probably because processing lower than 500 ROIs did not exceed the capacity of the mask R-CNN detectors [15,23].

4.8. Preening Behavior Measurement with the Trained Detector

Individual preening hens could be continuously monitored with the trained detector. The extracted behavior information showed that the hens showed temporal and spatial preference on the preening during the testing period. These behaviors may provide valuable insights into farm management and facility design. For example, hens may show displacement preening around feeders when they cannot access feed [25] and understanding the frequency of preening hens present around feeders may help to evaluate the sufficiency of feeder allowance. At the current stage, we just explored the probability of using deep learning to detect the preening behavioral responses and further research is recommended to determine the thresholds of the responses with regards to welfare evaluation. Overall, the mask R-CNN preening behavior detector is a useful tool to evaluate hen preening behaviors.

5. Conclusions

This study developed mask R-CNN preening behavior detectors by modifying the ResNets, pre-trained weights, image resizers and number of ROIs. The detectors accurately segmented individual preening hens (MIOU: 83.6–88.7%) and had decent performance on detecting preening and non-preening hens, in which the precision, recall, specificity, accuracy, F1 score and AP were mostly over 80%. The overall processing speed for preening detection ranged from 342.9 to 413.7 ms·image^{-1}. The ResNet101 performed better on the preening detection among the six ResNets. The pre-trained COCO weights had better transfer learning efficiency than the pre-trained ImageNet weights. The image resizers in the 'None,' 'Square' and 'Pad64' modes performed similarly on hen preening detection. The 30 ROIs had the highest precision and specificity but the lower recall and AP among various numbers of ROIs, while more than 100 ROIs had similar performance on hen preening detection. With the trained detector, temporospatial preening behaviors of individual hens could be extracted. Overall, the mask R-CNN preening behavior detector is a useful tool to detect hen preening behaviors.

Author Contributions: Conceptualization, G.L.; methodology, G.L. and F.L.; software, G.L.; validation, X.H. and G.L.; formal analysis, G.L.; investigation, G.L.; resources, G.L.; data curation, G.L.; writing—original draft preparation, G.L.; writing—review and editing, F.L. and Y.Z.; visualization, G.L.; supervision, Y.Z.; project administration, G.L.; funding acquisition, Y.Z. All authors have read and agreed to the published version of the manuscript.

Funding: This research was funded by Egg Industry Center (EIC) and supported by USDA National Institute of Food and Agriculture multistate project under accession number 1020090.

Acknowledgments: We appreciate assistance provided by USDA technicians during the experiment.

Conflicts of Interest: The authors declare no conflict of interest. The funders had no role in the design of the study; in the collection, analyses or interpretation of data; in the writing of the manuscript or in the decision to publish the results.

Nomenclature

AP	Average precision	MIOU	Mean intersection over union
ARS	Agriculture Research Service	MP	Max pooling
Bbox	Bounding box	NMS	Non-maximum suppression
Box1–Box5	Proposed boxes with various sales and ratios after the region proposal network	$P_{interp}(r)$	Interpolated precision in the precision-recall curve when recall is r
C1–C5	Convolutional stages 1 to 5 in the residual network	$P(\widetilde{r})$	Measured precision at recall $\widetilde{r}$
COCO	Common object in context	P2-P6	Feature maps in the feature pyramid network
Conv.	Convolution	ReLu	Rectified linear units
CNN	Convolutional neural network	ResNet	Residual network
Faster RCNN	Faster region-based convolutional neural network	ResNet18-ResNet1000	Residual network with 18-1000 layers of convolution
FC	Fully-connected	ROI	Regions of interest
FCN	Fully-connected network	RPN	Region proposal network
FN	False negative	TN	True negative
FP	False positive	TP	True positive
FPN	Feature pyramid network	Ups.	Upsampling
IOU	Intersection over union	USA	United State of American
Mask R-CNN	Mask region-based convolutional neural network	USDA	United State department of agriculture

References

1. Powers, R.; Li, N.; Gibson, C.; Irlbeck, E. Consumers' Evaluation of Animal Welfare Labels on Poultry Products. *J. Appl. Commun.* **2020**, *104*, 1a. [CrossRef]
2. Xin, H. Environmental challenges and opportunities with cage-free hen housing systems. In Proceedings of the XXV World's Poultry Congress, Beijing, China, 5–9 September 2016; pp. 5–9.
3. Webster, A.J. Farm animal welfare: The five freedoms and the free market. *Vet. J.* **2001**, *161*, 229–237. [CrossRef] [PubMed]
4. Appleby, M.C.; Mench, J.A.; Hughes, B.O. *Poultry Behaviour and Welfare*; CABI: Oxfordshire, UK, 2004.
5. Delius, J. Preening and associated comfort behavior in birds. *Ann. N. Y. Acad. Sci.* **1988**, *525*, 40–55. [CrossRef] [PubMed]
6. Kristensen, H.H.; Burgess, L.R.; Demmers, T.G.; Wathes, C.M. The preferences of laying hens for different concentrations of atmospheric ammonia. *Appl. Anim. Behav. Sci.* **2000**, *68*, 307–318. [CrossRef]
7. Dawkins, M.S. Time budgets in red junglefowl as a baseline for the assessment of welfare in domestic fowl. *Appl. Anim. Behav. Sci.* **1989**, *24*, 77–80. [CrossRef]
8. Duncan, I. Behavior and behavioral needs. *Poult. Sci.* **1998**, *77*, 1766–1772. [CrossRef]
9. Nicol, C. Social influences on the comfort behaviour of laying hens. *Appl. Anim. Behav. Sci.* **1989**, *22*, 75–81. [CrossRef]
10. Banerjee, D.; Biswas, S.; Daigle, C.; Siegford, J.M. Remote activity classification of hens using wireless body mounted sensors. In Proceedings of the 9th International Conference on Wearable and Implantable Body Sensor Networks, London, UK, 10–12 May 2012; pp. 107–112.
11. Li, G.; Li, B.; Shi, Z.; Zhao, Y.; Ma, H. Design and evaluation of a lighting preference test system for laying hens. *Comput. Electron. Agric.* **2018**, *147*, 118–125. [CrossRef]
12. Li, G.; Ji, B.; Li, B.; Shi, Z.; Zhao, Y.; Dou, Y.; Brocato, J. Assessment of layer pullet drinking behaviors under selectable light colors using convolutional neural network. *Comput. Electron. Agric.* **2020**, *172*, 105333. [CrossRef]
13. Tian, Y.; Yang, G.; Wang, Z.; Li, E.; Liang, Z. Instance segmentation of apple flowers using the improved Mask R-CNN model. *Biosys. Eng.* **2020**, *193*. [CrossRef]
14. Li, G.; Xu, Y.; Zhao, Y.; Du, Q.; Huang, Y. Evaluating convolutional neural networks for cage-free floor egg detection. *Sensors* **2020**, *20*, 332. [CrossRef] [PubMed]
15. He, K.; Gkioxari, G.; Dollár, P.; Girshick, R. Mask R-CNN. In Proceedings of the IEEE international conference on computer vision, Venice, Italy, 22–29 October 2017; pp. 2961–2969.
16. Li, D.; Chen, Y.; Zhang, K.; Li, Z. Mounting behaviour recognition for pigs based on deep learning. *Sensors* **2019**, *19*, 4924. [CrossRef] [PubMed]
17. Yu, Y.; Zhang, K.; Yang, L.; Zhang, D. Fruit detection for strawberry harvesting robot in non-structural environment based on Mask R-CNN. *Comput. Electron. Agric.* **2019**, *163*, 104846. [CrossRef]
18. Lin, T.-Y.; Maire, M.; Belongie, S.; Hays, J.; Perona, P.; Ramanan, D.; Dollár, P.; Zitnick, C.L. Microsoft coco: Common objects in context. In Proceedings of the European conference on computer vision, Zurich, Switzerland, 6–12 September 2014; pp. 740–755.
19. Deng, J.; Dong, W.; Socher, R.; Li, L.-J.; Li, K.; Fei-Fei, L. Imagenet: A large-scale hierarchical image database. In Proceedings of the IEEE conference on computer vision and pattern recognition, Miami, FL, USA, 20–25 June 2009; pp. 248–255.
20. Sharif Razavian, A.; Azizpour, H.; Sullivan, J.; Carlsson, S. CNN features off-the-shelf: An astounding baseline for recognition. In Proceedings of the IEEE conference on computer vision and pattern recognition workshops, Columbus, OH, USA, 24–27 June 2014; pp. 806–813.
21. Lotter, W.; Sorensen, G.; Cox, D. A multi-scale CNN and curriculum learning strategy for mammogram classification. In *Deep Learning in Medical Image Analysis and Multimodal Learning for Clinical Decision Support*; Springer: New York, NY, USA, 2017; pp. 169–177.
22. Abdulla, W. Mask R-CNN for Object Detection and Instance Segmentation on Keras and Tensorflow. Available online: https://github.com/matterport/Mask_RCNN (accessed on 30 March 2020).
23. Huang, J.; Rathod, V.; Sun, C.; Zhu, M.; Korattikara, A.; Fathi, A.; Fischer, I.; Wojna, Z.; Song, Y.; Guadarrama, S. Speed/accuracy trade-offs for modern convolutional object detectors. In Proceedings of the IEEE conference on computer vision and pattern recognition, Honolulu, HI, USA, 21–26 July 2017; pp. 7310–7311.

24. International Egg Commission. Atlas of the Global Egg Industry. Available online: https://www.internationalegg.com/wp-content/uploads/2015/08/atlas_2013_web.pdf (accessed on 16 June 2020).
25. Duncan, I.; Wood-Gush, D. An analysis of displacement preening in the domestic fowl. *Anim. Behav.* **1972**, *20*, 68–71. [CrossRef]
26. Koelkebeck, K.W.; Amoss, M., Jr.; Cain, J. Production, physiological and behavioral responses of laying hens in different management environments. *Poult. Sci.* **1987**, *66*, 397–407. [CrossRef] [PubMed]
27. Vezzoli, G.; Mullens, B.A.; Mench, J.A. Relationships between beak condition, preening behavior and ectoparasite infestation levels in laying hens. *Poult. Sci.* **2015**, *94*, 1997–2007. [CrossRef]
28. Goodfellow, I.; Bengio, Y.; Courville, A. *Deep Learning*; MIT Press: Cambridge, MA, USA, 2016.
29. He, K.; Zhang, X.; Ren, S.; Sun, J. Deep residual learning for image recognition. In Proceedings of the IEEE conference on computer vision and pattern recognition, Las Vegas, LA, USA, 27–30 June 2016; pp. 770–778.
30. Everingham, M.; Van Gool, L.; Williams, C.K.; Winn, J.; Zisserman, A. The pascal visual object classes (voc) challenge. *IJCV* **2010**, *88*, 303–338. [CrossRef]
31. Vala, H.J.; Baxi, A. A review on Otsu image segmentation algorithm. *Int. J. Adv. Res. Comput. Eng. Technol.* **2013**, *2*, 387–389.
32. Aydin, A.; Cangar, O.; Ozcan, S.E.; Bahr, C.; Berckmans, D. Application of a fully automatic analysis tool to assess the activity of broiler chickens with different gait scores. *Comput. Electron. Agric.* **2010**, *73*, 194–199. [CrossRef]
33. Chen, L.-C.; Papandreou, G.; Kokkinos, I.; Murphy, K.; Yuille, A.L. Semantic image segmentation with deep convolutional nets and fully connected crfs. *arXiv* **2014**, arXiv:1412.7062.
34. Lin, T.-Y.; Dollár, P.; Girshick, R.; He, K.; Hariharan, B.; Belongie, S. Feature pyramid networks for object detection. In Proceedings of the IEEE conference on computer vision and pattern recognition, Honolulu, HI, USA, 21–26 July 2017; pp. 2117–2125.
35. COCO. Detection Evaluation. Available online: http://cocodataset.org/#detection-eval (accessed on 15 April 2020).
36. Du, C.; Wang, Y.; Wang, C.; Shi, C.; Xiao, B. Selective feature connection mechanism: Concatenating multi-layer CNN features with a feature selector. *PaReL* **2020**, *129*, 108–114. [CrossRef]

animals

MDPI

Article

A Machine Vision-Based Method Optimized for Restoring Broiler Chicken Images Occluded by Feeding and Drinking Equipment

Yangyang Guo [1], Samuel E. Aggrey [1], Adelumola Oladeinde [1,2], Jasmine Johnson [1], Gregory Zock [1] and Lilong Chai [1,*]

1 Department of Poultry Science, University of Georgia, Athens, GA 30602, USA; yangyang.guo@uga.edu (Y.G.); saggrey@uga.edu (S.E.A.); ade.oladeinde@usda.gov (A.O.); jcj36120@uga.edu (J.J.); gregzock@uga.edu (G.Z.)

2 U.S. National Poultry Research Center, USDA-ARS, Athens, GA 30605, USA

* Correspondence: lchai@uga.edu

Simple Summary: The equipment in the poultry house can occlude top view images of broiler chickens and limit the efficiency of vision-based target detection. In this study, we sought to improve the efficiency of a previously developed method to detect and restore broiler chicken areas blocked by feeders and drinkers. To do this, we developed and tested linear and elliptical fitting restoration methods under different occlusion scenarios to restore occluded broiler chicken areas. The restoration method correctly restored the occluded broiler chicken area >80% of the time. This study provides a practical approach to enhancing the image quality in applying a machine vision-based method for monitoring poultry health and welfare.

Abstract: The presence equipment (e.g., water pipes, feed buckets, and other presence equipment, etc.) in the poultry house can occlude the areas of broiler chickens taken via top view. This can affect the analysis of chicken behaviors through a vision-based machine learning imaging method. In our previous study, we developed a machine vision-based method for monitoring the broiler chicken floor distribution, and here we processed and restored the areas of broiler chickens which were occluded by presence equipment. To verify the performance of the developed restoration method, a top-view video of broiler chickens was recorded in two research broiler houses (240 birds equally raised in 12 pens per house). First, a target detection algorithm was used to initially detect the target areas in each image, and then Hough transform and color features were used to remove the occlusion equipment in the detection result further. In poultry images, the broiler chicken occluded by equipment has either two areas (TA) or one area (OA). To reconstruct the occluded area of broiler chickens, the linear restoration method and the elliptical fitting restoration method were developed and tested. Three evaluation indices of the overlap rate (OR), false-positive rate (FPR), and false-negative rate (FNR) were used to evaluate the restoration method. From images collected on d2, d9, d16, and d23, about 100-sample images were selected for testing the proposed method. And then, around 80 high-quality broiler areas detected were further evaluated for occlusion restoration. According to the results, the average value of OR, FPR, and FNR for TA was 0.8150, 0.0032, and 0.1850, respectively. For OA, the average values of OR, FPR, and FNR were 0.8788, 0.2227, and 0.1212, respectively. The study provides a new method for restoring occluded chicken areas that can hamper the success of vision-based machine predictions.

Keywords: broiler chicken; machine vision; image restoring; precision poultry farming

Citation: Guo, Y.; Aggrey, S.E.; Oladeinde, A.; Johnson, J.; Zock, G.; Chai, L. A Machine Vision-Based Method Optimized for Restoring Broiler Chicken Images Occluded by Feeding and Drinking Equipment. *Animals* **2021**, *11*, 123. https://doi.org/10.3390/ani11010123

Received: 10 December 2020
Accepted: 6 January 2021
Published: 8 January 2021

1. Introduction

The computer or machine vision-based technology (MVT) has been suggested and tested to monitor livestock and poultry behaviors [1–4], health [5,6], and flock activity [7–9].

At present, the techniques used to obtain information about poultry include 3D vision technology [10,11], infrared thermal imaging technology [12,13], and image processing technology [14–16]. 3D vision technology can effectively obtain the target area through the three-dimensional information in the scene. Aydin [17] used 3D vision technology to detect broilers and assessed the lameness of broiler chickens. 3D vision is more time-consuming than 2D vision because of the larger amount of data in 3D. Infrared thermal imaging technology uses temperature information to remove interferences and obtain poultry target areas. Zaninelli et al. [18] built an animal monitoring system based on infrared imaging technology and pattern recognition to detect a hen in a closed room of a housing system. Thermal imaging of poultry surface temperature is not consistent so that the low-temperature area tends to be lost. Image processing technology distinguishes target and non-target areas through image information characteristics. We used image processing technology to detect the area of the broiler chicken in the video scene and further analyzed the distribution of the broiler chicken [19]. Although the target area could be detected through image processing technology, it was dependent on the information in the scene. When the scene changes, the detection method may not be effective. In poultry houses, the complex production systems, such as feeding and drinking equipment (e.g., water pipe, feeder, and hanging chains), is a critical challenge for collecting top view animal images because animals or poultry are occluded in the images, which leads to the high uncertainty in analyzing animal information (e.g., behaviors and body features).

The poultry body is similar to an ellipse, so many researchers used ellipse fitting to obtain poultry information. Lao et al. [20] used contour ellipse fitting to obtain ten behavioral parameters. Further, the Naive Bayes Classification method has been used to classify and distinguish six behaviors of preening, shaking, resting, wing flapping, exploration, and wing lifting. Amraei et al. [21] performed ellipse fitting on the body of the chicken to obtain relevant parameters and conducted weight estimation through artificial neural networks. Poursaberi et al. [22] extracted the boundary of the bird and the parameters of the best-fitted ellipse to categorize turning, wing flapping, lying, and standing behaviors. In addition, the research on detecting elliptical targets has also achieved some results. Liu et al. [23] proposed a fast and effective ellipse detection method, which performed better detection results. Dong et al. [24] combined the advantages of arc extraction and arc grouping to propose an ellipse detection method. Therefore, the ellipse fitting method is one of the suitable methods for broiler chicken target detection.

The aforementioned methods can be modified to remove image interferences with ellipse fitting and obtain relevant chicken movement information. The objectives of this study were (1) develop an imaging processing strategy for removing equipment and restore occluded chicken areas; (2) test the effect of the optimized method to remove equipment areas; (3) evaluate the efficiency of different image restoration methods used in this study for two primary occlusion scenarios.

2. Materials and Methods

2.1. Experimental Setup and Data Collection

This study was conducted in two identical experimental broiler facilities on the Poultry Research Farm at the University of Georgia, Athens, USA. Unless otherwise stated, the experimental setup and data have been previously published in [19]. Briefly, six identical pens (measuring 1.84 L $\times$ 1.16 W m, 20 Cobb 500 broiler chickens/pen, the density was about 0.11 m^2 floor per bird) were monitored separately with a high definition (HD) camera (PRO-1080MSFB, Swann Communications, Santa Fe Springs, CA, USA) mounted on the ceiling (2.5 m above floor) to capture video (15 frame/s with the resolution of 1440 $\times$ 1080 pixels). Video/image acquisition time was from 13 February 2020, to 18 March 2020. Collected videos were further analyzed and processed by MATLAB-R2019b (The MathWorks, Inc., Natick, MA USA).

2.2. *Method for Target Detection*

From our observation, the equipment interference in images of chickens became less with an increase in the birds' size and age. Therefore, the first four weeks of chicken images were selected as research samples in this study. The method for target detection has been developed and published; see our other paper [19].

Figure 1 shows the images collected on d2, d9, d16, and d23. The hanging feeder was installed when birds were two weeks old and were tall enough to use it. Thus, images of d2 and d9 have a floor feeder, and d16 and d23 have a hanging feeder.

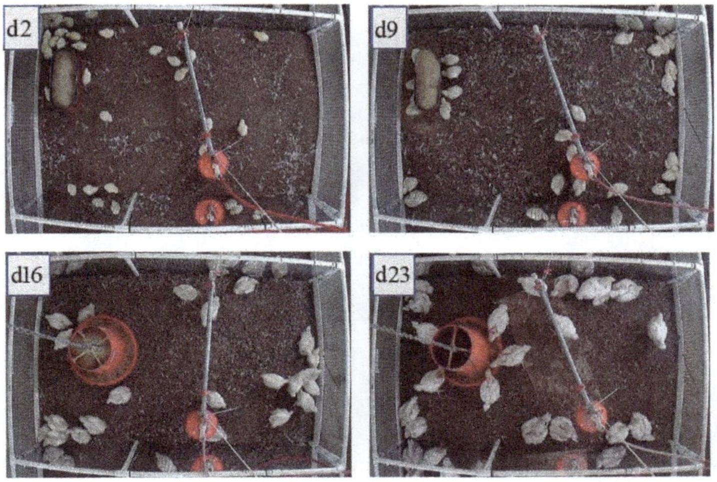

Figure 1. Examples of experimental data collection.

Figure 2 is an image collected on d23 as an example to show the target detection method. It can be seen from Figure 2c that the nipple drinker caused interference in birds' detection. Therefore, it is necessary to remove this interference to improve the quality of the chicken's images.

Figure 2. Examples of target detection results: (**a**) generation of binary image; (**b**) binary classification; and (**c**) the binary image obtained based on (**a**).

2.3. *The Equipment Area Removal*

According to pre-processing of images, we identified that image occlusion was caused by the presence of three pieces of equipment: (1) the water pipe; (2) the water pressure regulator (red circle area at the end of water pipe); and (3) the feeder. Therefore, the current study focused on the restoration of images occluded by the presence of three pieces of equipment.

(1) Water pipe interference removal.

The Hough transform can effectively detect the straight line in an image [25], so the method was used and modified to detect the pipe area in broiler houses. The Hough transform was performed on the image in Figure 2c to retain only the maximum peak (Figure 3a). Figure 3b,b′ show lines (green) that have passed the peak point with the 'yellow ×' indicating the starting point of the lines and the 'red ×' indicating the ending point of the lines. The first starting point in the 'yellow ×' was selected as the starting point of the pipe, and the last ending point in the 'red ×' was the ending point of the pipe. The connection line was approximately the centerline of the pipeline. Therefore, it was considered that the area obtained by the left and right extensions of 5 pixels based on the line was the pipe area, as shown in the red area in Figure 3c. Figure 3d shows the images with the pipe blocking area removed.

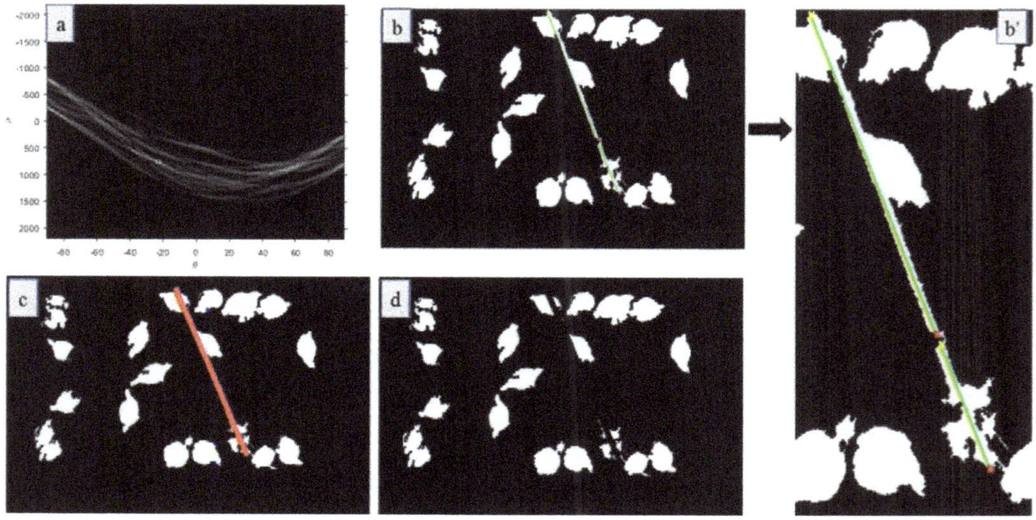

Figure 3. The process for pipe area detection and removal. The white box in (**a**) is the maximum peak point; In (**b**), it shows all the lines (green) that have passed the peak point with the yellow × indicating the starting point of the lines and the red × indicating the ending point of the lines. The drinking zone was zoomed-in (**b′**); In (**c**), the red area is the pipe area; and (**d**) is the result of pipe area removal.

(2) Water pressure regulator interference removing.

The color of the water pressure regulator (i.e., circular area at the end of the water pipe in Figure 1) was red, which was quite different from the color of the broiler chickens. Therefore, the circular area of the water pressure regulator was removed by color information to obtain chicken profiles in Figure 4. It can be seen from Figure 4 that part of the broiler chicken's missing area was caused by the occlusion of the water pipe, water pressure regulator, and feeder. For instance, the broiler chicken was divided into two areas in the yellow box in Figure 4.

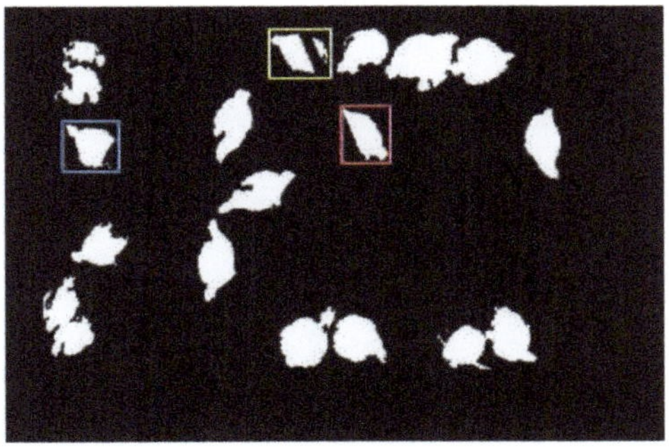

Figure 4. The result of equipment area removal. The area loss of the broiler chicken in the blue, yellow, and red boxes were caused by the feeder, water pipe, and water pressure regular, respectively.

2.4. Equipment Occlusion Detection and Restoration

Figure 5 shows the three equipment that occluded broiler chicken images, i.e., feeder (Figure 5a), water pipe (Figure 5b,c), and water pressure regulator (Figure 5d). There are two occlusion scenarios: (i) the body of chicken was divided into two areas (TA) (see Figure 5b) and (ii) the body of chicken was partly occluded, so the body has only one area (OA) (see Figure 5a,c,d). In this study, our method was modified to restore occlusion areas for both TA and OA scenarios.

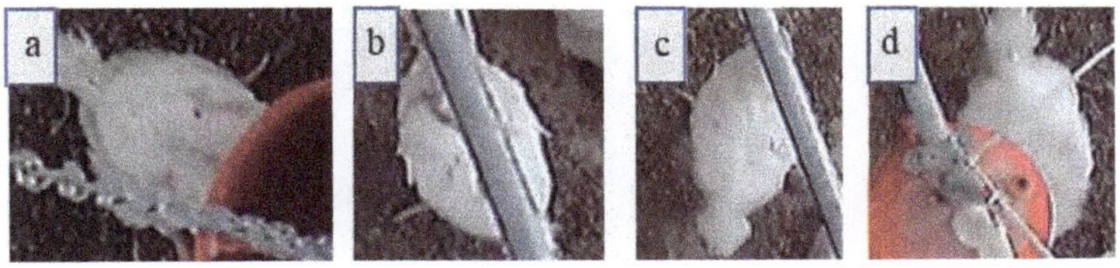

Figure 5. Examples of common equipment occlusions: (**a**) feeder occlusion; (**b**) water pipe occlusion–two areas (TA); (**c**) water pipe occlusion–one area (OA); (**d**) water pressure regulator occlusion.

(1) Image restoration for TA occlusion.

To restore images of broiler chickens in the drinking zone, we removed the image background by keeping the water pipe and its surrounding area (i.e., red rectangular box area in Figure 6a). We performed a Linear Morphological Closure Operation (Linear Restoration Method) on the red box along the direction of the vertical water pipe, as shown in Figure 6b.

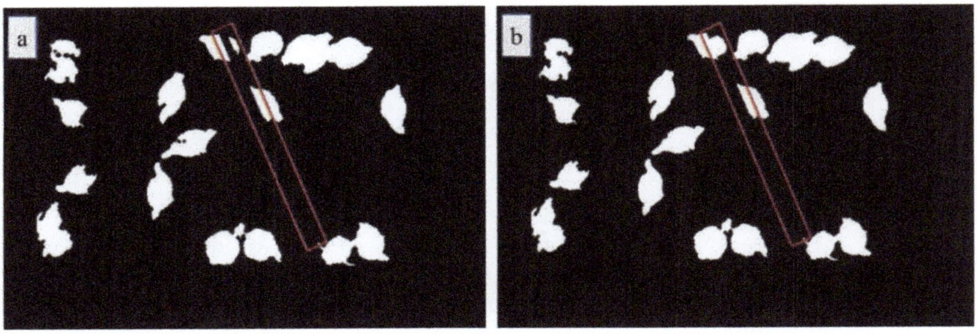

Figure 6. Occlusion area restoration for TA: (**a**) removed pipe and its surrounding area; (**b**) restored pipe area and chickens' images.

(2) Image restoration for OA occlusion.

Since the body shape of a broiler chicken is elliptical, the ellipse fitting method [26,27] was used to restore the occluded part of the broiler chicken. The ellipse fitting can be expressed using Equation (1).

$$a \times x^2 + b \times y^2 + c \times x + d \times y + e \times x \times y + f = 0 \quad (1)$$

$a \neq 0$, so Equation (2) can be changed to:

$$x^2 + \frac{b}{a} \times y^2 + \frac{c}{a} \times x + \frac{d}{a} \times y + \frac{e}{a} \times x \times y + \frac{f}{a} = 0 \quad (2)$$

where x, y are variables, x is the abscissa of images, y is the ordinate of images. a, b, c, d, e, f are constants.

Five coordinates are needed to determine the ellipse. In this paper, the ellipse was calculated and obtained from five points selected from the boundary of the unobstructed body of the broiler chickens. Figure 7 shows the example how occlusion area under OA situation was restored.

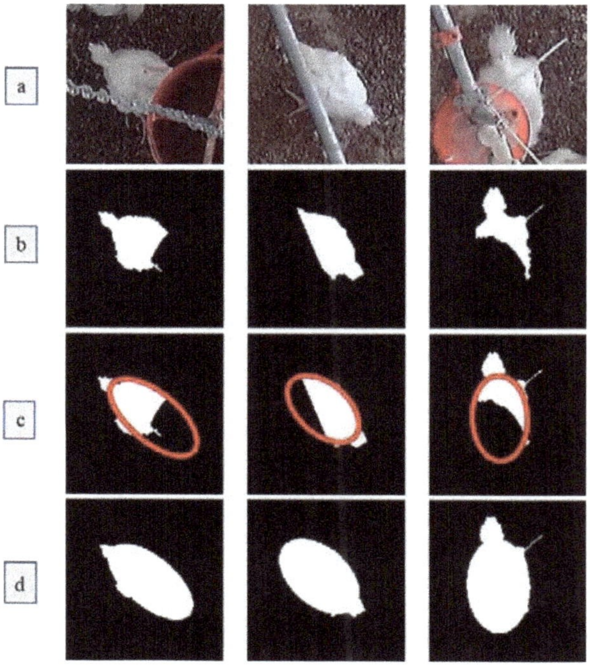

Figure 7. Example of the occlusion area restoration for OA: (**a**) is the original occlusion image; (**b**) is the corresponding detected binary image; (**c**) is the corresponding fitted ellipse (the red ellipse); and (**d**) is the restored result.

2.5. Evaluation Criteria and Statistical Analysis

Three evaluation indices were used to evaluate the image restoration methods: the overlap rate (OR), false-positive rate (FPR), and false-negative rate (FNR) [28].

The OR is the percentage of the actual target region affected by the overlapping of the actual target region and the restored target region. The higher the OR, the larger the overlap region and the better the restoration effect. The OR was calculated with Equation (3):

$$\mathrm{OR} = (N_1 \cap N_2)/N_1 \times 100\% \quad (3)$$

where N_1 is the real region indicated by artificial marking and N_2 is the region indicated by the restoration method.

The FPR is the percentage of the background region misjudged as the target region. The lower value, the better the restoration effect. The FPR was calculated with Equation (4):

$$FPR = [N_2 - (N_1 \cap N_2)]/N_1 \times 100\% \quad (4)$$

The FNR is the percentage of the target region misjudged as background. A lower value indicates a better effect of the restoration. The FNR was calculated with Equation (5):

$$FNR = [N_1 - (N_1 \cap N_2)]/N_1 \times 100\% \quad (5)$$

A one-way ANOVA (MATLAB-R2019b) was used to test if there were significant differences in OR, FPR, or FNR under different scenarios (e.g., one area and two areas) of occlusions. The effect was significant when the p-value was less than 0.05.

3. Results and Discussion

3.1. Restoration Efficiency for Occluded Area

About 100 images collected on d2, d9, d16, and d23 were selected for the new method evaluation. The restoration effect on occluded area of the chicken is shown in Figure 8, where red boxes are the broiler chickens occluded by the equipment. Basically, all images show occlusions, more or less.

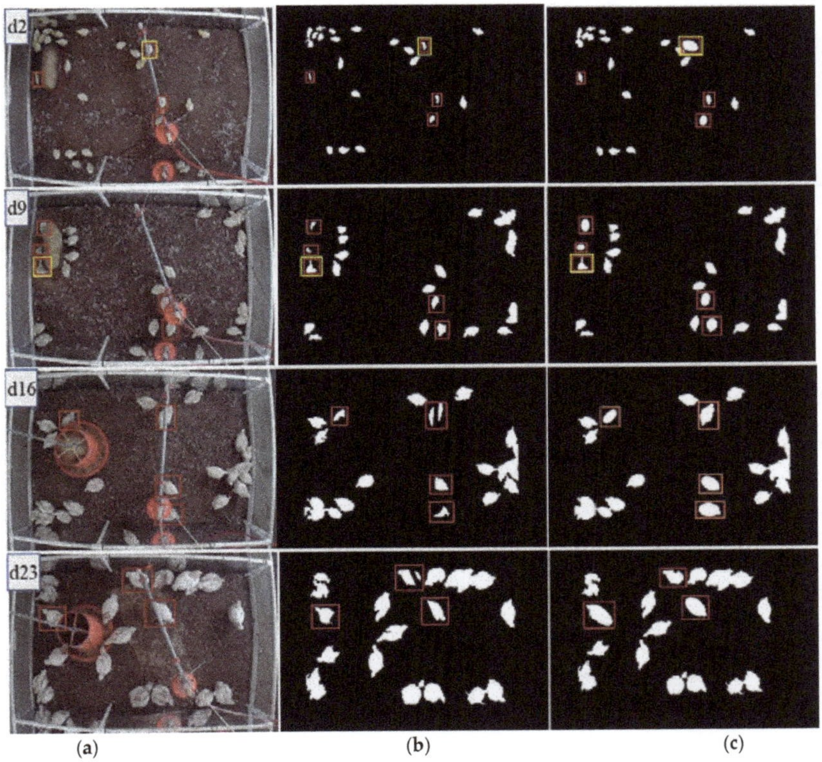

Figure 8. Restoration result of the occluded area. (**a**) original image; (**b**) the binary image of the broiler occluded area before restoration; (**c**) the binary image of the broiler occluded area after restoration. Red boxes are the broiler chickens occluded by the equipment. The yellow boxes are special cases, such as crowding.

It can be seen from Figure 8 that the broiler chicken area occluded by the equipment can be fixed. However, when the occlusion is extensive, such as the yellow box highlighted area (collected on d2) in Figure 8, the ellipse was overfitted. Therefore, we ascertained that it was not suitable to use ellipse fitting when chickens are crowded together (2 or more chickens in a group) and occluded by equipment (e.g., the yellow box of d9 in Figure 8 has two broiler chickens blocked by the feeder). In this study, we used ellipse fitting only when a single broiler chicken area was occluded.

To analyze the difference between the occluded area of chickens before and after image/occlusion restoration, the ratio of chicken area images before and after the restoration was quantified. Since the chicken posture was not uniform, the area of an individual chicken was determined by taking an average of the area of complete images of broiler chickens. From the d2, d9, d16, and d23 experimental images, about 150 broiler chicken area samples were obtained for TA and OA, respectively.

It can be seen from Table 1 that the linear restoration method can restore the occluded area well for the TA scenario, and there was a slight but not significant difference in the area compensation effect on different days/bird ages (p = 0.056). In this case, the occluded area of the pipe was relatively regular, so the restoration effect was superior. In the case of OA, there was no significant difference in the area compensation effect (p = 0.333) on different days/bird ages. We observed that the elliptical fitting restoration method can restore the occlusion area when the occlusion was not extensive better (e.g., when less than <50% of the broiler chicken area was occluded/blocked) (Figure 8; Table 1). When occlusion was extensive, ellipse underfitting or overfitting occurred (i.e., the restoration area was either too small or too large and likely contributed to the lack of significance in OA).

Table 1. The ratio of a broiler chicken area occluded by the equipment before and after restoration compared to an intact broiler chicken area (150 samples each).

Occlusion Type	d2		d9		d16		d23	
	BF_{area}/IN_{aera}	AF_{area}/IN_{aera}	BF_{area}/IN_{aera}	AF_{area}/IN_{aera}	BF_{area}/IN_{aera}	AF_{area}/IN_{aera}	BF_{area}/IN_{aera}	AF_{area}/IN_{aera}
TA	0.4357	0.9706	0.4971	0.9687	0.6299	0.9637	0.6747	0.9512
OA	0.4773	1.2008	0.4518	0.9445	0.5077	1.009	0.6962	1.1017

Note: BF_{area} is the average area of broiler chickens before restoration; AF_{area} is the average area of broiler chickens after restoration; IN_{aera} is the area of the intact broiler chicken area (not occluded by an equipment). TA—two areas; OA—one area.

3.2. *Performance of the Restoration Method*

When the occluded area of the broiler chicken could not be determined, predicting the actual area of the broiler chicken was not possible either. In the case of TA, the shape of the occluded area was regular (i.e., elliptical shape), so we could approximate the overall area of the broiler chicken as the actual area to evaluate the linear restoration method. Figure 9a shows the image with a bird blocked by water pipe (TA occlusion) and then reconstructed with the method developed in this study (i.e., ellipse fitting restoration).

In the case of OA, the occluded area of the broiler chicken was irregular, which made it difficult to obtain the actual area of the broiler chicken. Therefore, to determine the complete broiler chicken area, we artificially removed some parts of the area to simulate occlusion and performed ellipse fitting restoration on the removed area to evaluate the restoration efficiency (Figure 9b). We selected 80 suitable target images from d2, d9, d16, and d23 broiler chicken images to determine the average values of OR, FPR, and FNR (Table 2).

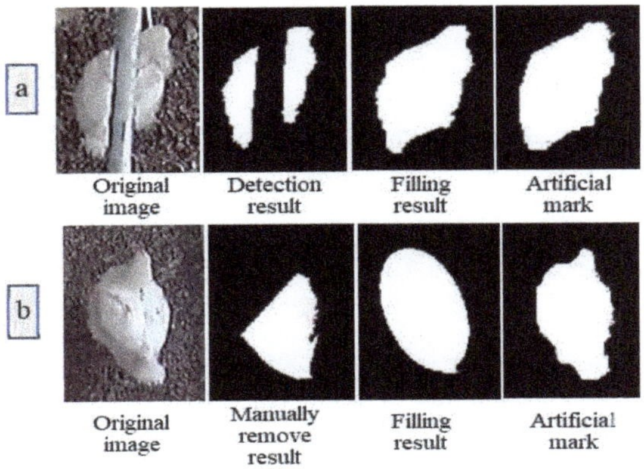

Figure 9. Comparison between restored chicken image area and an intact chicken image area (without occlusion): (**a**) process of image restoration for the TA situation; and (**b**) an intact chicken image without occlusion but artificially edited as the OA situation by removing a part for method evaluation.

Table 2. Comparison of average values on different days.

Evaluation Indices	Occlusion Type	d2	d9	d16	d23	Average
OR	TA	0.7265 [a]	0.8240 [b]	0.8593 [c]	0.8502 [c]	0.8150
	OA	0.9106 [a]	0.8480 [a]	0.8673 [a]	0.8834 [a]	0.8788
FPR	TA	0.0022 [a]	0.0076 [b]	0.0031 [c]	0.0002 [d]	0.0032
	OA	0.2963 [a]	0.2216 [ab]	0.2064 [ab]	0.1665 [b]	0.2227
FNR	TA	0.2735 [a]	0.1760 [b]	0.1407 [c]	0.1498 [c]	0.1850
	OA	0.0894 [a]	0.1520 [a]	0.1327 [a]	0.1166 [a]	0.1212

Note: In the same row, different letters of a, b, c and d represent significant differences among the means ($p \leq 0.05$); Evaluation indices include the overlap rate (OR), false-positive rate (FPR), and false-negative rate (FNR).

It can be seen from Table 2 that for TA, the average values of OR, FPR, and FNR were 0.8150, 0.0032, and 0.1850, respectively. The smaller OR value and the larger FNR value of d2 and d9 were significantly different from other days ($p < 0.05$) because the broiler chickens were small, and the occluded area was relatively large, resulting in a large area of the broiler being lost. For the OR value and the FNR values of d16 and d23, there were no differences ($p = 0.297$). From Table 2, it can be concluded that as the broiler grows, the difference in linear restoration results decreases.

For OA, the average values of OR, FPR, and FNR at different ages were 0.8788, 0.2227, and 0.1212, respectively. Since the broiler chicken area was not elliptical, the ellipse fitting restores would classify part of the background area as the target area resulting in a larger FPR value (0.2227). The larger OR value (0.9106) and the smaller FNR value (0.0894) for d2 images was because the broiler chickens were small. Picking different points on the body boundary resulted in different ellipses, which affected the restoration result, thereby leading to a lack of significance in OR in images on four different days ($p = 0.111$), in FPR in images of d2, d9, and d16 ($p = 0.082$), and FNR in images of four different days ($p = 0.111$).

In the current study, the selection of 5-points needs further improvement to optimize the ellipse fitting. In addition, monitoring individual poultry behaviors (e.g., feeding, drinking, lying, standing, walking, etc.) needs to be studied separately in occlusion restoration because behavior postures are different from each other. We developed a method based on the shape feature of the broiler chicken (ellipse) to restore the occluded area of the broiler chickens. In addition, some other machine learning or deep learning algorithms, such as support vector machines, have been reported with a similar function in image processing. Comparing ellipse fitting to other machine learning or deep learning algorithms/models is required to develop or optimize the method for occlusion restoration and other automatic methods for poultry behaviors or health monitoring.

4. Summary and Conclusions

In this study, a machine vision-based method was optimized to restore broiler chickens' images occluded by an equipment. According to the pre-processing of images, the general occlusion was identified as two area occlusion (TA) and one area (OA) occlusion. Three evaluation indices include the overlap rate (OR), false-positive rate (FPR), and false-negative rate (FNR), were used to evaluate the restoration method.

For TA occlusion, the average values of OR, FPR, and FNR were 0.8150, 0.0032, and 0.1850, respectively. The linear restoration effect was better than elliptical fitting, which was less affected by the growth of the broiler chicken because the occluded area was regular/normal in the case of TA. For OA occlusion, the average values of OR, FPR, and FNR were 0.8788, 0.2227, and 0.1212, respectively. The method we optimized/developed was not applicable for some special situations, such as crowding.

In the future study, the occluded area of crowded broiler chickens will be segmented first and then restored. In addition, monitoring individual poultry behaviors (e.g., feeding, drinking, lying, standing, walking, etc.) needs to be studied separately in occlusion removal because behavior postures are different from each other.

Author Contributions: L.C. and Y.G. came up with the research thoughts; Y.G., L.C., S.E.A. and A.O. figured out the research methodology; S.E.A. and A.O. designed the poultry biological experiment; Y.G., L.C., S.E.A., A.O. and J.J. managed research resources; Y.G. developed the algorithm for data analysis; S.E.A. and J.J. managed the chickens; Y.G., L.C., S.E.A., A.O., J.J. and G.Z. collected the original data; Y.G. and L.C. analyzed the data; Y.G. and L.C. verified the results; Y.G., L.C., S.E.A. and A.O. wrote the manuscript. All authors have read and agreed to the published version of the manuscript.

Funding: USDA-ARS cooperative grants for the University of Georgia (58-6040-6-030 and 58-6040-8-034). USDA Agricultural Research Service project number: 6040-32000-010-00-D; USDA-NIFA Hatch Project (GEO00895): Future Challenges in Animal Production Systems: Seeking Solutions through Focused Facilitation.

Institutional Review Board Statement: Not applicable.

Informed Consent Statement: Not applicable.

Data Availability Statement: Not applicable.

Acknowledgments: This study was partially supported by a cooperative grant 58-6040-6-030 (L.C.) and 58-6040-8-034 (S.E.A.) from the United State Department of Agriculture-Agriculture Research Service; USDA-NIFA Hatch Project (GEO00895): Future Challenges in Animal Production Systems: Seeking Solutions through Focused Facilitation. The funders have no role in the study design and implementation, data collection and analysis, decision to publish, or preparation of this manuscript. We thank Marie Milfort for her technical assistance.

Conflicts of Interest: The authors declare that they have no conflicts of interest.

References

1. Li, G.; Zhao, Y.; Purswell, J.L.; Du, Q.; Chesser, G.D., Jr.; Lowe, J.W. Analysis of feeding and drinking behaviors of group-reared broilers via image processing. *Comput. Electron. Agric.* **2020**, *175*, 105596. [CrossRef]
2. Pereira, D.F.; Miyamoto, B.C.; Maia, G.D.; Sales, G.T.; Magalhães, M.M.; Gates, R.S. Machine vision to identify broiler breeder behavior. *Comput. Electron. Agric.* **2013**, *99*, 194–199. [CrossRef]
3. Wang, J.; Wang, N.; Li, L.; Ren, Z. Real-time behavior detection and judgment of egg breeders based on YOLO v3. *Neural Comput. Appl.* **2020**, *32*, 5471–5481. [CrossRef]
4. Li, G.; Ji, B.; Li, B.; Shi, Z.; Zhao, Y.; Dou, Y.; Brocato, J. Assessment of layer pullet drinking behaviors under selectable light colors using convolutional neural network. *Comput. Electron. Agric.* **2020**, *172*, 105333. [CrossRef]
5. Zhuang, X.; Zhang, T. Detection of sick broilers by digital image processing and deep learning. *Biosyst. Eng.* **2019**, *179*, 106–116. [CrossRef]

6. Okinda, C.; Lu, M.; Liu, L.; Nyalala, I.; Muneri, C.; Wang, J.; Zhang, H.; Shen, M. A machine vision system for early detection and prediction of sick birds: A broiler chicken model. *Biosyst. Eng.* **2019**, *188*, 229–242. [CrossRef]
7. Neves, D.P.; Mehdizadeh, S.A.; Tscharke, M.; de Alencar Nääs, I.; Banhazi, T.M. Detection of flock movement and behaviour of broiler chickens at different feeders using image analysis. *Inf. Process. Agric.* **2015**, *2*, 177–182. [CrossRef]
8. Fernandez, A.P.; Norton, T.; Tullo, E.; van Hertem, T.; Youssef, A.; Exadaktylos, V.; Vranken, E.; Guarino, M.; Berckmans, D. Real-time monitoring of broiler flock's welfare status using camera-based technology. *Biosyst. Eng.* **2018**, *173*, 103–114. [CrossRef]
9. Lee, D.H.; Kim, A.K.; Choi, C.H.; Kim, Y.J. Study on image-based flock density evaluation of broiler chicks. *J. Korean Inst. Inf. Electron. Commun. Technol.* **2019**, *12*, 373–379.
10. Mortensen, A.K.; Lisouski, P.; Ahrendt, P. Weight prediction of broiler chickens using 3D computer vision. *Comput. Electron. Agric.* **2016**, *123*, 319–326. [CrossRef]
11. Nakarmi, A.D.; Tang, L.; Xin, H. Automated tracking and behavior quantification of laying hens using 3D computer vision and radio frequency identification technologies. *Trans. ASABE* **2014**, *57*, 1455–1472.
12. Nääs, I.A.; Garcia, R.G.; Caldara, F.R. Infrared thermal image for assessing animal health and welfare. *JABB-Online Submiss. Syst.* **2014**, *2*, 66–72. [CrossRef]
13. Xiong, X.; Lu, M.; Yang, W.; Duan, G.; Yuan, Q.; Shen, M.; Norton, T.; Berckmans, D. An Automatic Head Surface Temperature Extraction Method for Top-View Thermal Image with Individual Broiler. *Sensors* **2019**, *19*, 5286. [CrossRef]
14. Barbin, D.F.; Mastelini, S.M.; Barbon Jr, S.; Campos, G.F.; Barbon, A.P.A.; Shimokomaki, M. Digital image analyses as an alternative tool for chicken quality assessment. *Biosyst. Eng.* **2016**, *144*, 85–93. [CrossRef]
15. Mehdizadeh, S.A.; Neves, D.P.; Tscharke, M.; Nääs, I.A.; Banhazi, T.M. Image analysis method to evaluate beak and head motion of broiler chickens during feeding. *Comput. Electron. Agric.* **2015**, *114*, 88–95. [CrossRef]
16. Xiao, L.; Ding, K.; Gao, Y.; Rao, X. Behavior-induced health condition monitoring of caged chickens using binocular vision. *Comput. Electron. Agric.* **2019**, *156*, 254–262. [CrossRef]
17. Aydin, A. Using 3D vision camera system to automatically assess the level of inactivity in broiler chickens. *Comput. Electron. Agric.* **2017**, *135*, 4–10. [CrossRef]
18. Zaninelli, M.; Redaelli, V.; Luzi, F.; Bontempo, V.; Dell'Orto, V.; Savoini, G. A monitoring system for laying hens that uses a detection sensor based on infrared technology and image pattern recognition. *Sensors* **2017**, *17*, 1195. [CrossRef]
19. Guo, Y.; Chai, L.; Aggrey, S.E.; Oladeinde, A.; Johnson, J.; Zock, G. A Machine Vision-Based Method for Monitoring Broiler Chicken Floor Distribution. *Sensors* **2020**, *20*, 3179. [CrossRef]
20. Lao, F.; Teng, G.; Li, J.; Yu, L.; Li, Z. Behavior recognition method for individual laying hen based on computer vision. *Trans. Chin. Soc. Agric. Eng.* **2012**, *28*, 157–163.
21. Amraei, S.; Abdanan Mehdizadeh, S.; Salari, S. Broiler weight estimation based on machine vision and artificial neural network. *Br. Poult. Sci.* **2017**, *58*, 200–205. [CrossRef]
22. Poursaberi, A.; Wichman, A.; Bahr, C.; Laura, H.Ã.; Pastell, M.; Berckmans, D. Automatic monitoring of turkeys: A vision-based approach to detect and analyse the behaviour of turkeys in transport cages based on ellipse fitting. In Proceedings of the 7th World Congress on Computers in Agriculture Conference Proceedings, Reno, NV, USA, 22–24 June 2009.
23. Liu, Z.; Liu, X.; Duan, G.; Tan, J. A real-time and precise ellipse detector via edge screening and aggregation. *Mach. Vision. Appl.* **2020**, *31*, 1–23. [CrossRef]
24. Dong, H.; Prasad, D.K.; Chen, I.M. Accurate detection of ellipses with false detection control at video rates using a gradient analysis. *Pattern Recogn.* **2018**, *81*, 112–130. [CrossRef]
25. Aggarwal, N.; Karl, W.C. Line detection in images through regularized Hough transform. *IEEE Trans. Image Process.* **2006**, *15*, 582–591. [CrossRef]
26. Fitzgibbon, A.; Pilu, M.; Fisher, R.B. Direct least square fitting of ellipses. *IEEE Trans. Pattern Anal.* **1999**, *21*, 476–480. [CrossRef]
27. Rosin, P.L. Further five-point fit ellipse fitting. *Graph. Model. Image Process.* **1999**, *61*, 245–259. [CrossRef]
28. Guo, Y.; He, D.; Song, H. Region detection of lesion area of knee based on colour edge detection and bilateral projection. *Biosyst. Eng.* **2018**, *173*, 19–31. [CrossRef]

animals

Article

Smart Feeding Unit for Measuring the Pecking Force in Farmed Broilers

Rogério Torres Seber [1], Daniella Jorge de Moura [1], Nilsa Duarte da Silva Lima [2] and Irenilza de Alencar Nääs [1,2,*]

1 School of Agricultural Engineering, University of Campinas, Campinas, Av. Cândido Rondon, 501 Barão Geraldo, São Paulo 13083-875, Brazil; rogerio.torres.seber@gmail.com (R.T.S.); djmoura@unicamp.br (D.J.d.M.)

2 Graduate Program in Production Engineering, Paulista University, São Paulo 04026-002, Brazil; nilsa.lima@stricto.unip.br

* Correspondence: irenilza@unicamp.br

Simple Summary: We present a novel method for assessing broiler pecking force data during feeding. The prototype consisted of a power supply unit with a data acquisition module, management software connected to a computer for data storage, and a video camera to verify the pecking force during signal processing. The acquisition, processing, and classification of the pecking force signal information were valuable during broilers' feeding. The smart feeding unit (SFU) prototype was useful in the continuous generation of information that could be applied to evaluate the amount of pecking force and performance during the broilers' growth.

Abstract: Feeding is one of the most critical processes in the broiler production cycle. A feeder can collect data of force signals and continuously transform it into information about birds' feed intake and quickly permit more agile and more precise decision-making concerning the broiler farm's production process. A smart feeding unit (SFU) prototype was developed to evaluate the broiler pecking force and average feed intake per pecking (g/min). The prototype consisted of a power supply unit with a data acquisition module, management software connected to a computer for data storage, and a video camera to verify the pecking force during signal processing. In the present study, seven male Cobb-500 broilers were raised in an experimental chamber to test and commission the prototype. The prototype consisted of a feeding unit (feeder) with a data acquisition module (amplifier), with real-time integration for testing and intuitive operation with Catman Easy software connected to a computer to obtain and store data from signals. The sampling of average feed intake per pecking per broiler (g) was conducted during the first minute of feeding, subtracting the amount of feed provided per the amount of feed consumed, including the count of pecking in the first minute of feeding. An equation was used for estimating the average feed intake per pecking per broiler (g). The results showed that the average broiler pecking force was 1.39 N, with a minimum value of 0.04 N and a maximum value of 7.29 N. The average feed intake per pecking (FIP) was 0.13 g, with an average of 173 peckings per minute. The acquisition, processing, and classification of signals in the pecking force information were valuable during broilers' feeding. The smart feeding unit prototype for broilers was efficient in the continuous assessment of feed intake and can generate information for estimating broiler performance.

Keywords: broiler; feeding system; pecking force; precision livestock farming

Citation: Seber, R.T.; Moura, D.J.d.; Lima, N.D.d.S.; Nääs, I.d.A. Smart Feeding Unit for Measuring the Pecking Force in Farmed Broilers. *Animals* **2021**, *11*, 864. https://doi.org/10.3390/ani11030864

Academic Editors: Yang Zhao, María Cambra-López, Daniella Jorge De Moura, Weichao Zheng and Maria Caria

Received: 11 January 2021
Accepted: 14 March 2021
Published: 18 March 2021

1. Introduction

Broiler production contributes significantly to Brazilian agribusiness. In 2019, Brazilian broiler meat production volume was 13,245 metric tons, corresponding to 13.4% of the world market (98,594 metric tons). The poultry industry is evolving to meet the global demand for animal protein with low environmental impact [1–3], with integrated, vertical production, and applying technologies in the production process. Such actions aim to increase the productive efficiency index [4,5], improve the welfare and health of animals [6,7] with lower production costs without compromising parameters of welfare, performance, and quality [8]. These initiatives also lead to improving the consumer perception of broiler meat [9]. The use of technology is essential to manage modern broiler farms, and providing relevant information to farmers enhances their decision-making during the production cycle.

Intelligent equipment with a rapid response in real-time has been extensively studied and used in the poultry production process. Besides contributing to precision livestock farm development, these studies apply connectivity and analysis tools in real-time and use algorithms to monitor the production cycle during animal growth, behavior, welfare, health aspects, and performance [10–18]. Current literature presents developed equipment for poultry farming using sounds, images, and modeling of signals of force and pecking sounds to monitor animals' growth and other factors related to the welfare, behavior, and feeding of birds [11,16,19–23]. Although the use of an automatic recording of sounds for animal husbandry and health management (quantitative analysis) in other species, such as swine, show similar results in automatic and manual assessments of the frequency of coughing, the disadvantage of manual assessments is the time spent compared to that of automatic assessments [24].

Animal feeding is one of the essential processes during broiler production and one of the most studied subjects with methods of analyzing sound signals and video images. Previous studies aimed to assess broilers' food consumption using scales, including pecking sounds and developed a pecking classification algorithm for continuous and non-invasive broiler production monitoring [21,22]. Integrating the previously studied variables, a pecking detection system including video footage, a microphone to record sound signals, a scale to record bird weight automatically, and the use of a group pecking classification algorithm were used to evaluate the short-term broiler feeding behavior [11]. An automated system (group-housed individual turkey feeder and bodyweight measurement station) was developed to monitor the turkeys' feeding and body weight in real-time. The monitoring system consisted of hardware and software subsystems (hardware subsystem: mechanical framework of feed stations, radio frequency identification components, electronic scales, communication modules, and a central computer; software subsystem: a hardware monitoring and data acquisition program, and a data processing and management program). The system was tested with a group of turkeys to assess data on the frequency of feeding behavior and performance [25]. Another study using signal analysis methods developed a chicken pecking force equation on an automatic feeder. The equation involves mathematical and statistical approaches to analyze the chicken pecking force at different stages of production. The pecking force was related to parameters of feed flow rate and more accurate decision-making regarding the hopper aperture in the feeder at the production process [23]. The results indicated that the birds' satiety level determined the pecking force described by a polynomial function [23].

The coupled cranial kinesis (the ability to move the upper beak relative to the braincase [26]) in domestic fowl does not play a dominant role in the feeding process. The jaw drops just after lifting the upper jaw, suggesting that the coupled cranial kinesis does not necessarily happen while the food is grasped [16]. Similar traits may occur in the following cycles for the food going into the oral cavity during the feeding process. However, the coupled kinesis is applied when the bird closes the beak since it cannot depress the upper jaw without raising the lower jaw [27]. However, the bird can adapt specific beak movement depending on the type of food, and such behaviors are subordinate to the constraints of the beak morphological structure [16]. Foraging is a natural bird behavior, and during foraging, it also pecks the ground, and often broilers pecking does not result in the retention of a feed particle [28]. Therefore, we need to continuously check the head and beak movements to assess feed behavior and consequent performance.

In the current literature, we did not find a study that has directly measured the signs of broiler head movement to identify and classify the pecking force during feeding that was specially instrumented for monitoring broiler feeding behavior. We believe that in the future, such equipment associated with the signal interpretation may provide us with a unique ability to manage production data regarding feed performance and detection of numbers of birds per feeder in a non-intrusive way. An intelligent feeding unit can collect data from force signals and transform it into information such as the pecking force during the feed intake. Such continuous information such as feed intake per pecking, activity, number of birds around the feeder, and weight gain by feed intake allows faster and more accurate decision-making regarding the farm level's production process. Therefore, the present study aimed to develop a prototype of an intelligent feeding unit (smart feeding unit, SFU) to evaluate and register the broilers' pecking force during feeding by correlating the force applied with the actual catch of feed particles.

2. Materials and Methods

The development of the SFU consisted of constructing a prototype and commissioning the prototype for the acquisition of broilers' pecking force data.

2.1. The Prototype of the Smart Feeding Unit

Figure 1 presents the structure of the feeding unit. The parts that bring up the SFU are a 200 mm diameter feeder plate with a 20 mm height, a load cell (manufacturer Hottinger Baldwin Messtechnik—HBM), a base plate, and a fastener screw. The prototype was built using modular steel components. The feeding unit's height was adjusted according to the bird's height (as it grows). The equipment was installed on concrete support to maintain proper stability.

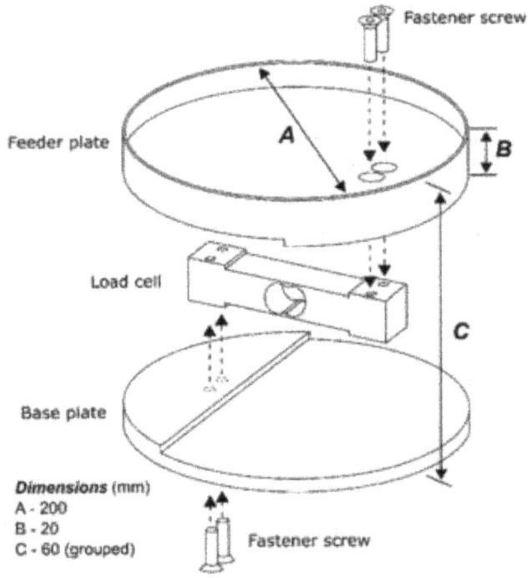

Figure 1. Schematic of the smart feeding unit prototype structure.

The prototype was subjected to tests during the adaptation phase of the birds to the new feeder. The prototype test started in the last week of the production cycle (35–41 days old), and validation was performed on the last day of the production cycle, lasting 24 h to collect signals from the birds' pecking force.

2.2. Experiment

Seven male Cobb-500 broilers were reared from 1 to 42 days old, considering the first five weeks (1 to 35 days old) as an adaptation phase to the environment, and in the last week, the test of the feed unit prototype (35 to 42 days old). From 35 days of age on, an instrumented feeder was made available to measure the pecking force during feeding (through a week before slaughter). Seven broilers were housed in an experimental floor chamber (experimental environmental controlled chamber) equipped with a tubular feeder (for the adaptation phase of the broiler feed), a pendant drinker, temperature sensors, air humidity control, electrical heater (for the initial rearing phase), air exhaust fan, mechanical cooling, and dimmable LED lighting (artificial light). These devices make the flow of heat supply and removal, vapor supply and removal, illumination, environmental sensing and control, and video monitoring.

The floor was covered with 5 cm thick wood shavings litter, which was reloaded whenever necessary. The daily lighting was 16 h during the growth period. Feeding and water were provided ad libitum during the experiment.

The experiment was carried out at the Animal Environmental Laboratory at the School of Agricultural Engineering at the University of Campinas (Unicamp, Brazil), and the study was approved by the University's Animal Ethics Committee (protocol number 5278-1/2019—CEUA—Unicamp).

2.3. Data Acquisition and Signal Processing

The data acquisition and signal processing consisted of the data acquisition module (QuantumX—MX840A amplifier, manufacturer Hottinger Baldwin Messtechnik—HBM), with real-time integration for testing and Catman Easy software (CatmanEasy version 4.2, manufacturer Hottinger Baldwin Messtechnik—HBM), connected to a computer for obtaining and storing data from signals (Figure 2). Data acquisition is the process of obtaining via sampling a signal from a sensor and converting it to an electrical value (usually a voltage level) and later conversion to a digital value for further computer processing. Moreover, sensors are the devices that convert one type of electrical or mechanical signal (input-signal) into another (output-signal), usually an electrical signal. Signal conditioning is a step of data acquisition that combines the signal emitted by the sensor installed in the feeder (input-signal), amplifying, filtering the noise, and converting an analog signal into a digital signal (output-signal), with the input in a computer. The signal amplification is the increasing signal for processing (or digitization) that can increase the signal input resolution or increase the signal-to-noise ratio. In signal conditioning, the frequency spectrum is filtered only to include the valid data and block any noise [29]. A video camera (Sharp Corporation, 470 lines with 3.6 mm converging lens) was utilized for acquiring the images for checking of pecking during signal processing for data analysis (synchronizing images and signals), maintained in continuous monitoring mode (Figure 2). The video images were synchronized with the acquisition of signals to validate when birds pecked to determine the average feed consumption per pecking.

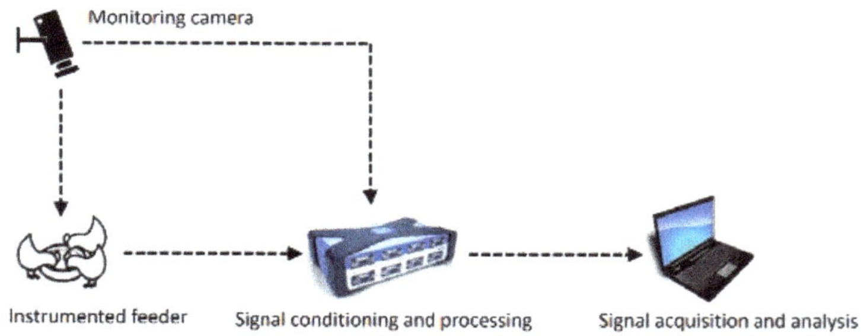

Figure 2. General scheme of signal conditioning and processing for data acquisition.

The software used to acquire, visualize, analyze, and report signal measurement data was Catman Easy (CatmanEasy version 4.2, manufacturer Hottinger Baldwin Messtechnik—HBM) to ensure the synchronism between the feed pecking image and the force signal acquisition of the pecked feeder. The signal analysis aims to extract information from the data acquired to generate some desirable information [30]. The feed pecking signals data reports during all the broilers feeding were registered and sent automatically to a computer. It was also possible to visualize the signal using graphical output.

2.4. Classification of Pecking and Validation

The automated real-time smart feeding unit prototype was tested in the last week of the broiler rearing cycle for system validation. The data collected by the data acquisition system contained raw data (with noises generated during the birds' movement and feeding) that were processed, organized, and filtered to remove the noise and later classify the signals of feed pecking and other eventual beak movements in the feeder. Video analysis was used as a tool in the validation of feed pecking (pecking vs. other beak movements). Image and signal force were synchronized for the pecking validation, ensuring that the feed was eaten. All force sampling occurred synchronically with the head-movement image acquisition, confirming that the registered pecking force originated from an observed feeding. The sampling selected was 1 min during bird feeding (sample size $n = 284$). Feed intake was estimated during a sampling synchronized image and signal force. The feed intake was calculated by subtracting the amount of feed provided (excluding the feeder's weight) per the amount of feed consumed, including the count of pecking in the one-minute concerning feeding intake.

Feed ration consumption was automatically assessed by the sensor system installed in the feeder and processed using the data acquisition system installed in the computer (Figure 2). The average feed intake per pecking per broiler (g) was estimated using Equation (1), adapted from Aydin et al. [21].

$$FIP\ (g) = \frac{TFI(g)}{TNP}, \qquad (1)$$

where *FIP* is feed intake per pecking, *TFI* is total feed intake (g), and *TNP* is total number of peckings per minute (*TNP*).

2.5. Statistical Analysis

The automated real-time smart feeding unit prototype was tested in the last week of the broiler rearing cycle for validation. The data collected by the data acquisition system contained raw data (including noises generated during the birds' movement and feeding) that were processed, organized, and filtered to remove this noise and later classify the signals (pecking vs. non-pecking). The comparison of the accurate feed pecking patterns and other noises were classified to define feeding behavior.

A table was generated from the data acquisition software containing the filtered and standardized data for the descriptive analysis and the *t*-test for a sample (one-way *t*-test) [31,32]. A one-way test is a hypothesis test that counts the chance of results only in one direction [31,33].

The descriptive analysis and *t*-test were applied to a sample size of 284 pecking force, measured in Newton (N). The average pecking force (6.5 N) of young chickens (<8 weeks old) tested in a poultry feeder (smaller hopper aperture) was utilized for the alternative hypothesis in the *t*-test [23]. The hypothesis test was presented, H0: all means are equal, vs. H1: at least one mean is different [32]. The data were analyzed using PAST software [34] (Paleontological Statistics version 4.03).

Sample tests are used to determine whether a single sample comes from a population with a given hypothetical average (alternative hypothesis μ_0). The alternative hypothesis is the data from the birds' pecking force (single sample) are equal to the mean of pecking force found in a previous study [23]. For the *t*-test of a sample (parametric), the confidence interval was 95% for the difference in means based on the standard error for estimating the mean and t distribution. The *t*-test statistic was calculated as expressed in Equation (2) [31]:

$$t = \frac{\overline{x} - \mu_0}{\frac{S}{\sqrt{n}}} \qquad (2)$$

where μ_0 is the hypothesized population mean, $\overline{x}$ is the sample mean, n is the sample size (number of observations), and S = is the sample standard deviation. Under the null hypothesis, the test statistic has Student's t distribution with $n-1$ degree of freedom.

3. Results

The test of the smart feeding unit (SFU) prototype allowed the acquisition of the pecking force data instantly during the broiler feeding. Figure 3 shows the pecking force data acquisition in one minute during the broilers' feeding, and Figure 4 shows the pecking force in ten seconds. The intensity and speed of data collection are due to the equipment's sensitivity in detecting the force when the broiler feeds.

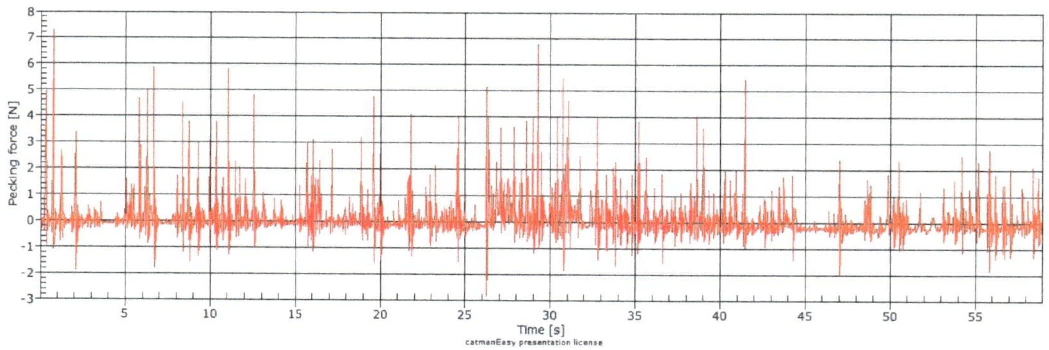

Figure 3. Pecking force at one-minute time length.

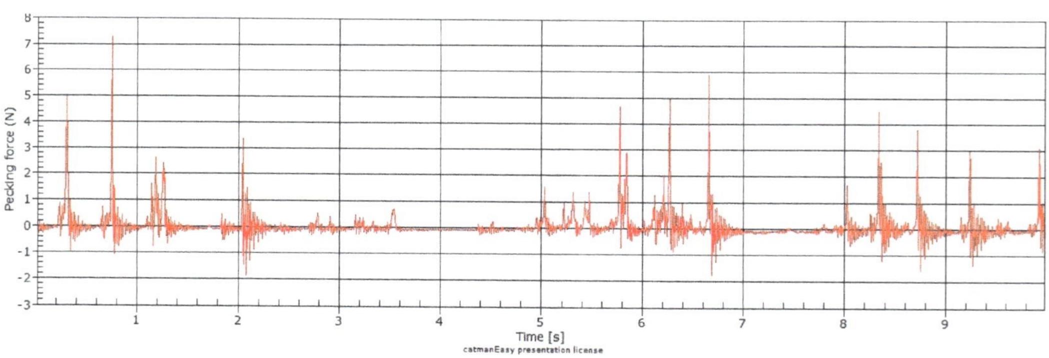

Figure 4. Pecking force at ten-second time length.

We observed a pecking force pattern during broiler feeding in the three images (Figures 3–5). Figure 6 represents the pecking force in 120 ms (milliseconds) described as one peck, showing the force variation applied to the feeder's sensor. The average feed intake per pecking (*FIP*) was 0.13 g, with an average of 173 peckings per minute (pecking frequency).

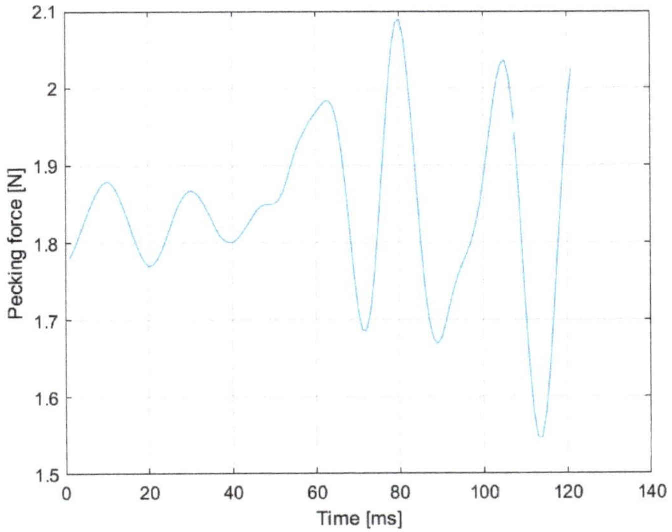

Figure 5. Pecking force at 120 ms.

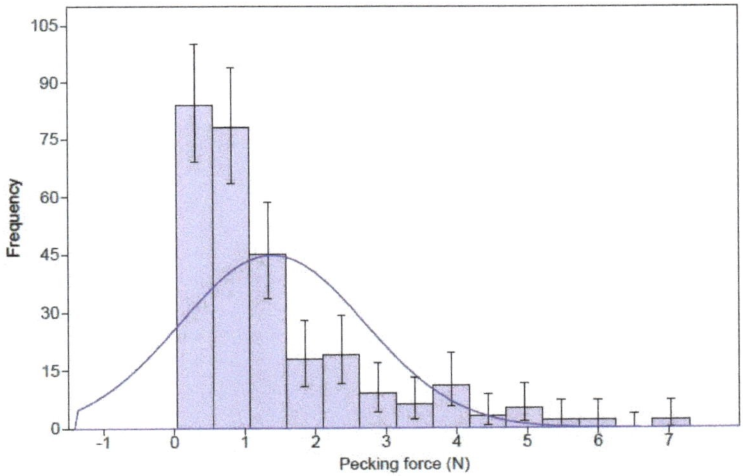

Figure 6. Frequency histogram of pecking force values.

The frequency distribution (histogram) with a normal fit adjusted in the histogram (Figure 6) shows a tilt of the tail to the right due to the birds' peak pecking force.

The results of descriptive statistics (Table 1) show that the average broiler pecking force was 1.39 N, with a minimum value of 0.04 N and a maximum value of 7.29 N from the analysis of 284 samples. The results of the *t*-test indicate that the mean (1.39 ± 0.15 N) of the broilers' pecking force differs ($p < 0.001$) from the average birds' pecking force in general (Table 1).

Table 1. Summary of results of descriptive analysis and *t*-test of broilers' pecking force.

Statistical Summary		Lower Confidence Limit	Upper Confidence Limit
n	284	284	284
Minimum	0.04	-	-
Maximum	7.29	-	-
Mean	1.39	1.24	1.54
Standard error	0.08	0.07	0.09
Standard deviation	1.31	1.15	1.49

One-sample *t*-test—Pecking force

t Statistic	*p*-value (2-tailed)	df [1]	Mean difference	95% Confidence interval of the difference	
				Lower	Upper
−65.66	<0.0001	283	5.11	1.24	1.54

[1] df: the degrees of freedom for the test, df = $n - 1$.

The classification of force signals in pecking was performed efficiently with the aid of synchronized video images to identify an effective pecking in a group or individual.

4. Discussion

The results showed that the broilers' average pecking force was validated by the smart feeding unit (SFU) prototype designed for fast response, with efficient readings for measuring force signals. We also observed a cyclic pattern of pecking force during broilers feeding that can be explained by the phases of the food intake process, specifically related to the biomechanical movement of the broiler's head [16,35]. A previous study evaluated broilers' biomechanical movement during feeding through computational analysis of images classifying the sequences of frames and kinematic variables to analyze the biomechanical behavior [35]. The authors considered six mandibulations (two sequences of three mandibulations) that involve a cycle of beak opening and a beak closing cycle. The studied kinematic variables included the head's displacement, the speed, and acceleration of beak opening and closing. Their results indicated that the birds' feeding behavior is divided into two phases, an appetite phase (exploration in search of feed) and a phase of actual feed consumption. Another relevant result was that birds are selective about the food's particle size in the initial phase of the production process. The authors concluded that birds' biomechanical patterns are related to different types of feed [16].

Another study carried out with hens of different ages (<8 weeks, 8–5 weeks, and >52 weeks old—weeks old birds), using an automatic feeder with different openings (hopper aperture: smaller, intermediate, and larger), developed a pecking force equation that was related to the feeding and hopper aperture of the feeder. The results showed that the birds' satiety level determined the pecking force. The maximum pecking force of chickens younger than eight weeks of age was 10 N after 40 min of feeding for the largest hopper opening. The pecking force was lower in the smaller feeder. The amount of feed consumed also decreased with the feeding time, indicating the birds' satiety leads to a lower pecking force [23]. The maximum broiler pecking force found was 7.29 N. Considering there are differences in feeder type, and the subject was broiler and not chicken, the average force for chickens, regardless of age, differed from those of the present study.

Other studies have also analyzed methods for assessing the broiler feed process [11,21,22]. However, the authors used sound signals as the primary method of detecting the broiler pecking. Pecking sounds from 10 broilers of 39-day-old were assessed while feeding, associated with video image records. The feeding system also registered the birds' weight simultaneously, and from there, they developed an algorithm to detect the birds pecking in groups. The feed intake was estimated automatically after the classification of the individual pecking sound that was detected by the algorithm. However, the proposed system did not exceed 90% accuracy in detecting broilers' pecking sounds when they were in group feeding because of the overlapping pecking sounds. The correlation was high considering the feed intake from broilers' pecking sound analysis. The authors concluded that a non-invasive and automated continuous system for monitoring chickens' feeding behavior could be essential for monitoring the growth process [22]. A previous study by the same group developed an algorithm for detecting broiler pecking to classify individual pecking and monitor feed intake. The pecking was accurately classified (93%), and feed consumption was correctly monitored (90%) from the sound analysis [21]. When analyzing the broilers' feeding behavior in the short term at a group level, including a system of evaluation by sound, image, and weighing monitored in real-time, the authors found a positive correlation between the used methods. The estimated precision was 90% when analyzing the meal size, 95% when evaluating the meal duration, 94% when studying the number of meals per day, and 89% for the feeding rate of broilers at 39 days of age from the analysis of the pecking sound and the bird weighing in an instrumented feeder [11].

In turkey production, an automated feed consumption and body weight monitoring system was also evaluated to assess feeding behavior. The automated system was based on ethernet and multithreading programming for real-time data acquisition. The authors [25] concluded that the system was effective in the acquisition and management of raw data and in the extraction of information on feeding behavior, which included the distribution of feeds over time, feed conversion rate at different stages of growth, pecking force, deglutition intervals, and meal breaks during feeding during turkey rearing.

Aydin et al. [21], using sound analysis, found that the 28-day-old broiler average feed intake per pecking was 0.025 g, and the average number of peckings was 85/min per broiler. However, the present study was conducted at 42 days of age and different strains. The average feed intake per pecking was 0.13 g, and the average number of peckings was 173/min per broiler. These results indicate that the equipment was fit to monitor broiler feeding. Broilers' age might explain the higher feed intake and the larger amount of pecking since intake increases according to the growth stage [36]. Further studies are needed to assess the relationship between feed intake and pecking force during feeding and at all stages of rearing.

The integration of smart sensors and technologies in the broiler rearing processor helps producers to optimize and minimize production losses and improves monitoring of birds in real-time [37–39]. The present study provided valuable information about an automatic broiler feeding system that collaborates with the application and implementation of intelligent systems as a data management tool during the broiler production process, potentially contributing to precision livestock farming to monitor the welfare of birds.

5. Conclusions

The smart feeding unit (SFU) as a broiler feeder has been tested and validated for its application to measure the bird's pecking force. The acquisition, processing, and classification of signals in information on the pecking force were valuable during broilers' feeding. The results confirmed that the average broiler pecking force was 1.39 ± 0.15 N and the average feed intake per pecking was 0.13 g, with an average of 173 peckings per minute.

This equipment can generate information that can serve as a basis for further studies on the performance, behavior, welfare of broilers, and automation of rearing processes regarding broiler feed management and can be easily adapted and included in other systems already on the farm.

Author Contributions: Conceptualization, R.T.S., D.J.d.M., and I.d.A.N.; methodology, R.T.S. and I.d.A.N.; formal analysis, R.T.S. and N.D.d.S.L.; investigation, R.T.S.; resources, R.T.S.; writing—original draft preparation, R.T.S., and N.D.d.S.L.; writing—review and editing, I.d.A.N. and N.D.d.S.L. This study is part of R.T.S. Ph.D. thesis advised by D.J.d.M. and I.d.A.N. All authors have read and agreed to the published version of the manuscript.

Funding: This study was financed in part by the Coordination of Improvement of Higher Education Personnel (CAPES, Brazil)—Finance Code 001.

Institutional Review Board Statement: The study was conducted according to the guidelines of the National Council for the Control of Animal Experimentation (CONCEA) and approved by the University's Animal Ethics Committee (protocol number 5278-1/2019, CEUA-Unicamp) of Animal Environmental Laboratory at the School of Agricultural Engineering at the University of Campinas (Unicamp, Brazil).

Data Availability Statement: Data will be available upon request to the first author.

Acknowledgments: The authors thank CAPES—Coordination for the Improvement of Higher Education Personnel (process number 001/2017) and CNPq—National Council for Scientific and Technological Development (process number 159842/2019-0) for the scholarships.

Conflicts of Interest: The authors declare no conflict of interest.

References

1. Mogensen, L.; Hermansen, J.E.; Halberg, N.; Dalgaard, R.; Vis, J.C.; Smith, B.G. Life Cycle Assessment Across the Food Supply Chain. In *Sustainability in the Food Industry*; Baldwin, C.J., Ed.; Iowa University: Ames, IA, USA, 2009; pp. 115–144, ISBN 978-0-8138-0846-8.
2. Röös, E.; Sundberg, C.; Tidåker, P.; Strid, I.; Hansson, P.-A. Can Carbon Footprint Serve as an Indicator of the Environmental Impact of Meat Production? *Ecol. Indic.* **2013**, *24*, 573–581. [CrossRef]
3. Leinonen, I.; Kyriazakis, I. How Can We Improve the Environmental Sustainability of Poultry Production? *Proc. Nutr. Soc.* **2016**, *75*, 265–273. [CrossRef]
4. Meluzzi, A.; Sirri, F. Welfare of Broiler Chickens. *Ital. J. Anim. Sci.* **2009**, *8*, 161–173. [CrossRef]
5. Brevik, E.; Lauen, A.Ø.; Rolke, M.C.B.; Fagerholt, K.; Hansen, J.R. Optimisation of the Broiler Production Supply Chain. *Int. J. Prod. Res.* **2020**, *58*, 5218–5237. [CrossRef]
6. Tuyttens, F.; Heyndrickx, M.; De Boeck, M.; Moreels, A.; Van Nuffel, A.; Van Poucke, E.; Van Coillie, E.; Van Dongen, S.; Lens, L. Broiler Chicken Health, Welfare and Fluctuating Asymmetry in Organic versus Conventional Production Systems. *Livest. Sci.* **2008**, *113*, 123–132. [CrossRef]
7. de Jong, I.C.; Hindle, V.A.; Butterworth, A.; Engel, B.; Ferrari, P.; Gunnink, H.; Perez Moya, T.; Tuyttens, F.A.M.; van Reenen, C.G. Simplifying the Welfare Quality® Assessment Protocol for Broiler Chicken Welfare. *Animal* **2016**, *10*, 117–127. [CrossRef] [PubMed]
8. Gocsik, É.; Brooshooft, S.D.; de Jong, I.C.; Saatkamp, H.W. Cost-Efficiency of Animal Welfare in Broiler Production Systems: A Pilot Study Using the Welfare Quality® Assessment Protocol. *Agric. Syst.* **2016**, *146*, 55–69. [CrossRef]
9. Pouta, E.; Heikkilä, J.; Forsman-Hugg, S.; Isoniemi, M.; Mäkelä, J. Consumer Choice of Broiler Meat: The Effects of Country of Origin and Production Methods. *Food Qual. Prefer.* **2010**, *21*, 539–546. [CrossRef]
10. Youssef, A.; Exadaktylos, V.; Berckmans, D.A. Towards Real-Time Control of Chicken Activity in a Ventilated Chamber. *Biosyst. Eng.* **2015**, *135*, 31–43. [CrossRef]
11. Aydin, A.; Berckmans, D. Using Sound Technology to Automatically Detect the Short-Term Feeding Behaviours of Broiler Chickens. *Comput. Electron. Agric.* **2016**, *121*, 25–31. [CrossRef]
12. Milosevic, B.; Ciric, S.; Lalic, N.; Milanovic, V.; Savic, Z.; Omerovic, I.; Doskovic, V.; Djordjevic, S.; Andjusic, L. Machine Learning Application in Growth and Health Prediction of Broiler Chickens. *Worlds Poult. Sci. J.* **2019**, *75*, 401–410. [CrossRef]
13. Wurtz, K.; Camerlink, I.; D'Eath, R.B.; Fernández, A.P.; Norton, T.; Steibel, J.; Siegford, J. Recording Behaviour of Indoor-Housed Farm Animals Automatically Using Machine Vision Technology: A Systematic Review. *PLoS ONE* **2019**, *14*, e0226669. [CrossRef]
14. Valletta, J.J.; Torney, C.; Kings, M.; Thornton, A.; Madden, J. Applications of Machine Learning in Animal Behaviour Studies. *Anim. Behav.* **2017**, *124*, 203–220. [CrossRef]
15. Zhuang, X.; Bi, M.; Guo, J.; Wu, S.; Zhang, T. Development of an Early Warning Algorithm to Detect Sick Broilers. *Comput. Electron. Agric.* **2018**, *144*, 102–113. [CrossRef]
16. Mehdizadeh, A.S.; Neves, D.P.; Tscharke, M.; Nääs, I.A.; Banhazi, T.M. Image Analysis Method to Evaluate Beak and Head Motion of Broiler Chickens during Feeding. *Comput. Electron. Agric.* **2015**, *114*, 88–95. [CrossRef]
17. Mortensen, A.K.; Lisouski, P.; Ahrendt, P. Weight Prediction of Broiler Chickens Using 3D Computer Vision. *Comput. Electron. Agric.* **2016**, *123*, 319–326. [CrossRef]
18. Amraei, S.; Abdanan Mehdizadeh, S.; Sallary, S. Application of Computer Vision and Support Vector Regression for Weight Prediction of Live Broiler Chicken. *Eng. Agric. Environ. Food* **2017**, *10*, 266–271. [CrossRef]
19. Fontana, I.; Tullo, E.; Butterworth, A.; Guarino, M. An Innovative Approach to Predict the Growth in Intensive Poultry Farming. *Comput. Electron. Agric.* **2015**, *119*, 178–183. [CrossRef]
20. Fontana, I.; Tullo, E.; Carpentier, L.; Berckmans, D.; Butterworth, A.; Vranken, E.; Norton, T.; Berckmans, D.; Guarino, M. Sound Analysis to Model Weight of Broiler Chickens. *Poult. Sci.* **2017**, *96*, 3938–3943. [CrossRef]

21. Aydin, A.; Bahr, C.; Viazzi, S.; Exadaktylos, V.; Buyse, J.; Berckmans, D. A Novel Method to Automatically Measure the Feed Intake of Broiler Chickens by Sound Technology. *Comput. Electron. Agric.* **2014**, *101*, 17–23. [CrossRef]
22. Aydin, A.; Bahr, C.; Berckmans, D. A Real-Time Monitoring Tool to Automatically Measure the Feed Intakes of Multiple Broiler Chickens by Sound Analysis. *Comput. Electron. Agric.* **2015**, *114*, 1–6. [CrossRef]
23. Asoegwu, S.N.; Obasi, A.U.; Ohanyere, S.O.; Nwakuba, N.R. Analytical Development of Bird Pecking Force Equation for a Self-Metering Poultry Feeder. *Agric. Eng. Int. CIGR J.* **2017**, *19*, 183–188.
24. Pessoa, J.; Rodrigues da Costa, M.; García Manzanilla, E.; Norton, T.; McAloon, C.; Boyle, L. Managing Respiratory Disease in Finisher Pigs: Combining Quantitative Assessments of Clinical Signs and the Prevalence of Lung Lesions at Slaughter. *Prev. Vet. Med.* **2021**, *186*, 105208. [CrossRef] [PubMed]
25. Tu, X.; Du, S.; Tang, L.; Xin, H.; Wood, B. A Real-Time Automated System for Monitoring Individual Feed Intake and Body Weight of Group Housed Turkeys. *Comput. Electron. Agric.* **2011**, *75*, 313–320. [CrossRef]
26. Bout, R.G.; Zweers, G.A. The Role of Cranial Kinesis in Birds. *Comp. Biochem. Physiol. A Mol. Integr. Physiol.* **2001**, *131*, 197–205. [CrossRef]
27. Van Den Heuvel, W.F. Kinetics of the Skull in the Chicken (Gallus Gallus Domesticus). *Neth. J. Zool.* **1991**, *42*, 561–582. [CrossRef]
28. Yo, T.; Vilariño, M.; Faure, J.M.; Picard, M. Feed Pecking in Young Chickens: New Techniques of Evaluation. *Physiol. Behav.* **1997**, *61*, 803–810. [CrossRef]
29. HBM, H.B.M. Sensor and Data Acquisition Articles from HBM. Available online: https://www.hbm.com/en/7329/technology-blog/ (accessed on 28 August 2020).
30. Sircar, P. *Mathematical Aspects of Signal Processing*; Cambridge University Press: Cambridge, UK, 2016; ISBN 978-1-107-17517-4.
31. Devore, J.L. Tests of Hypotheses Based on a Single Sample. In *Probability and Statistics for Engineering and the Sciences*; Cengage Learning: Boston, MA, USA, 2016; pp. 310–360, ISBN 978-1-305-25180-9.
32. Rosner, B. Hypothesis Testing: One-Sample Inference. In *Fundamentals of Biostatistics*; Cengage Learning: Boston, MA, USA, 2016; pp. 211–268. ISBN 978-1-305-26892-0.
33. Bruce, P.; Bruce, A. *Practical Statistics for Data Scientists: 50+ Essential Concepts*, 1st ed.; O'Reilly Media: Sebastopol, CA, USA, 2017; ISBN 978-1-4919-5296-2.
34. Hammer, Ø.; Harper, D.A.T.; Ryan, P.D. PAST: Paleontological Statistics Software Package for Education and Data Analysis. *Palaeontol. Electron.* **2001**, *4*, 9.
35. Neves, D.; Banhazi, T.; Nääs, I. Feeding Behaviour of Broiler Chickens: A Review on the Biomechanical Characteristics. *Braz. J. Poult. Sci.* **2014**, *16*, 1–16. [CrossRef]
36. Cobb-500 Cobb-500. In *Suplemento de Nutrição e Desempenho do Frango de Corte*; Cobb Vantress Brasil: Guapiaçu, Brasil, 2020.
37. Corkery, G.; Ward, S.; Kenny, C.; Hemmingway, P. Incorporating Smart Sensing Technologies into the Poultry Industry. *J. Worlds Poult. Res.* **2013**, *3*, 106–128.
38. Guo, Y.; Chai, L.; Aggrey, S.E.; Oladeinde, A.; Johnson, J.; Zock, G. A Machine Vision-Based Method for Monitoring Broiler Chicken Floor Distribution. *Sensors* **2020**, *20*, 3179. [CrossRef] [PubMed]
39. Li, N.; Ren, Z.; Li, D.; Zeng, L. Review: Automated Techniques for Monitoring the Behaviour and Welfare of Broilers and Laying Hens: Towards the Goal of Precision Livestock Farming. *Animal* **2020**, *14*, 617–625. [CrossRef] [PubMed]

animals

MDPI

Article

Design and Implementation of Poultry Farming Information Management System Based on Cloud Database

Haikun Zheng [1], Tiemin Zhang [1,2,3,*], Cheng Fang [1], Jiayuan Zeng [1] and Xiuli Yang [1]

[1] College of Engineering, South China Agricultural University, 483 Wushan Road, Guangzhou 510642, China; haikun0619@gmail.com (H.Z.); gu5457111@gmail.com (C.F.); zsural@126.com (J.Z.); xlyscau@scau.edu.cn (X.Y.)
[2] National Engineering Research Center for Breeding Swine Industry, Guangzhou 510642, China
[3] Guangdong Laboratory for Lingnan Modern Agriculture, Guangzhou 510642, China
* Correspondence: tm-zhang@163.com

Simple Summary: Informatization can effectively improve the production and management efficiency in the poultry farming process. In this study, a management system was designed to realize the acquisition, transmission, storage, and management of information, and upload the data to the cloud database to increase the flexibility and scalability of the system. On the basis of realizing production management functions, the system also incorporates an office management module, thus forming a complete data chain in production activities, so as to conduct farming data mining and accurate traceability in the next stage of the work. In particular, the system also adds poultry disease detection module supports to achieve the purpose of healthy farming. The research provides an information management plan for the intensive poultry farming model, and the designed management system may be the starting point of a future intelligent poultry farming management system based on cloud services and big data technology.

Abstract: Aiming at breaking down the bottleneck problems of different scale of poultry farms, the low profitability of poultry farming, and backward information management in China, a safe and efficient information management system for poultry farming was designed. This system consists of (1) a management system application layer, (2) a data service layer, and (3) an information sensing layer. The information sensing layer obtains and uploads production and farming information through the wireless sensor network built in the poultry house. The use of a cloud database as an information storage carrier in the data service layer eliminates the complex status of deploying local server clusters, and it improves the flexibility and scalability of the system. The management system application layer contains many sub-function modules including poultry disease detection functions to realize the visual management of farming information and health farming; each module operates independently and cooperates with each other to form a set of information management system for poultry farming with wide functional coverage, high service efficiency, safety, and convenience. The system prototype has been tested for the performance of wireless sensor network and cloud database, and the results show that the prototype is capable of acquiring and managing poultry farming information.

Keywords: poultry farming; information management; cloud database; disease detection

Citation: Zheng, H.; Zhang, T.; Fang, C.; Zeng, J.; Yang, X. Design and Implementation of Poultry Farming Information Management System Based on Cloud Database. *Animals* **2021**, *11*, 900. https://doi.org/10.3390/ani11030900

Academic Editors: Yang Zhao, Weichao Zheng, María Cambra-López and Daniella Jorge De Moura

Received: 2 February 2021
Accepted: 19 March 2021
Published: 22 March 2021

1. Introduction

Modern poultry farming companies need a complete management system to assist companies in managing their daily production activities. The system should cover such things as personnel office management, purchase–sales–inventory management, environmental monitoring and control in poultry houses, and monitoring of individual poultry information. At the same time, it also needs to include traceability management of products, diagnosis, and early warning of poultry diseases to meet the need for future development.

With the development of large-scale and intensive poultry farming, more intelligent and automated technologies and methods have been applied in poultry farming [1,2], such as radio frequency identification technology [3], Internet of Things technology (IoT), and cloud technology [4]. At the same time, there are methods such as poultry disease detection, poultry diet monitoring [5], environmental monitoring in poultry houses [6,7], product tracking and traceability, and abnormal detection in poultry houses to achieve precision farming. Yu Ligen et al. developed a network-based data acquisition system using LabVIEW software for environmental monitoring in poultry management [8], which describes the construction of data acquisition system hardware and the process of data acquisition. The method also provides a reference for us to build an environmental monitoring module. British Irvine explored the British broiler meat value chain [9] and provided a method for constructing the traceability module in the poultry farming management system through its in-depth analysis of the value chain. Research on applying wireless sensing systems along with mobile networking and cloud platform to some agricultural systems in crops [10] has provided us with new ideas to develop a similar information system for poultry farming.

In recent years, more and more researchers have devoted themselves to the study of precision poultry farming [11,12]. Some researchers help farmers control and monitor the health status of poultry through IOT, imagery analytics, and other technologies [13]. Other researchers build online platforms and using smart sensors to record and manage production information in real time [14–16]. Although wireless sensing and cloud platform techniques are well advanced, there is no complete system that covers all of the functions to meet needs for poultry farming management. The technical difficulties include the unified construction of the system, the reasonable division of functional modules, the good mutual cooperation between modules, the interaction of software and hardware, and the intelligentization of the system.

This paper reports the conceptual design of a poultry farming information management system with a cloud database as the core hub, through the connection of the underlying hardware facilities in the poultry house and the upper management system and the cloud database to manage the daily office work and production management tasks of poultry farming enterprises. In addition, this cloud-based management system also pays more attention to the storage and management of data information by separating the database system from the software system. The ultimate goal of the poultry farming information management system is to expand the development of the poultry industry management system with big data analysis capability.

2. Overall System Architecture

Figure 1 shows the overall structure of the system. The system is divided into three layers, which are the upper management system, the intermediate data service layer (also known as the middle layer), and the underlying layer (also known as the bottom layer, including hardware facilities in the poultry house).

The upper layer is a software management system that provides a good visual interface. The management system is divided into an office automation module, a production management module, an expert system, and a traceability module. As for the middle layer, the cloud database is used to store the data and information generated by the upper layer and the bottom layer, and at the same time construct a reasonable network environment to solve mutual communication problems by configuring the underlying server. At the bottom layer, in poultry house(s), environmental sensors, Wi-Fi receiving and transmitting devices, and single-chip microcomputers can be configured to timely acquire and transmit environmental information and poultry individual information (including poultry weight information, feed intake data information, drinking water data information, poultry egg quality information, etc.). Ventilation fans, evaporative cooling pads, heaters, and other equipment placed in the poultry house to regulate environmental parameters such as temperature and humidity in the house.

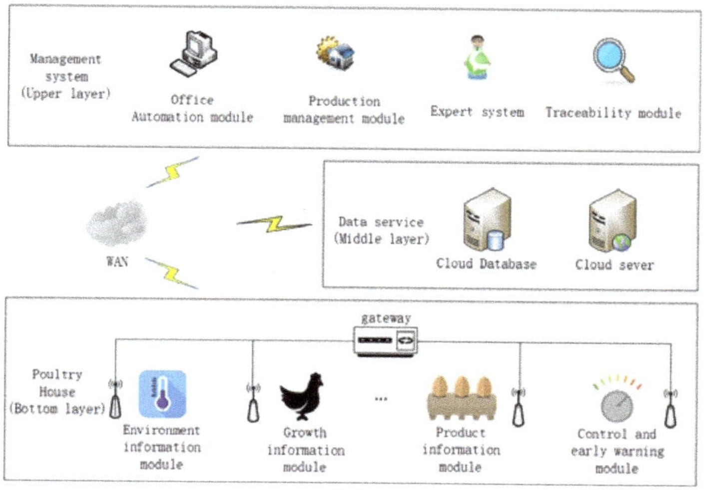

Figure 1. Overall structure of the system.

2.1. System Network Construction and Transmission Method

An environmental parameter information sensor (including temperature and humidity sensor, ammonia sensor, carbon dioxide sensor, hydrogen sulfide sensor, light intensity sensor, etc.), feed intake data monitoring module, drinking water data monitoring module, video monitoring system, fans, evaporative cooling pads, heaters, feeding, and manure cleaning facilities in the poultry house together form the local area network system in the house. This section mainly reports the information transmission methods, local transmission strategies, and configuration of network nodes in poultry houses.

Data transmission between the poultry house and the house environmental control system is primarily provided by a suitable wired communication system, such as a fieldbus. There are some disadvantages (e.g., configure too many network endpoints, device address assignment rules, and other issues) of using a full wireless system. Therefore, the local area network in the house uses a wireless/wireline hybrid construction [17]. As shown in Table 1, among the three commonly used wireless transmission modes (i.e., Bluetooth, Wi-Fi, and ZigBee), the Wi-Fi technology has the longest transmission distance and the fastest transmission speed [18]; therefore, Wi-Fi technology was selected as the wireless transmission method in the poultry house selects.

Table 1. Comparison of three commonly wireless transmission methods (Bluetooth, Wi-Fi, ZigBee) within four parameters (frequency band, transmission distance, power dissipation, transmission rate).

Transmission Modes	Frequency Band	Transmission Distance	Power Dissipation	Transmission Rate
Bluetooth	2.4 GHz	2–30 m	20 mA	1 Mbps
Wi-Fi	2.4 GHz	100–300 m	10–50 mA	600 Mbps
ZigBee	2.4 GHz	50–300 m	5 mA	100 Kbps

2.2. Cloud Database

The Alibaba Cloud Database RDS service was used in the system, as it has a good visual operation interface and numerous auxiliary analysis tools. It can generate database-related files such as E-R diagrams (Entity Relationship Diagram) and data dictionary with one click, and it can also generate test data to ensure the test works during database development. The database uses a relational database, and the database version is MySQL5.7. As the core hub of a poultry farming information management system, the cloud database should carry out requirements analysis, concept design, logical structure design, construction of the E-R model, design table structure, and primary-foreign key relationships in the process of design and construction.

The intranet address of the system can be accessed by the Alibaba Cloud Server, which has the advantages of fast reading speed and convenient setup. The external network address can be accessed by Internet users with access rights, and the database can be read and written. In the design stage, the selected database memory is 1024 MB, 1 core CPU, the storage capacity is 20 GB, and the maximum number of connections is 2000. It is confirmed in the actual development test that this configuration can meet the development needs.

2.3. Upper Management System

The management system uses the C++ language as the main development tool, the latest Qt5 framework as an open source support library, and the Qt Creator as an IDE (Integrated Development Environment) for compilation and development.

Figure 2 shows the functional framework of the poultry farming management system. The whole system is divided into four functional modules. The production management module mainly realizes the monitoring of environmental parameters in the poultry house, the monitoring of the growth information of individual poultry, and the management of the production operations in the poultry house. The office management module mainly fulfills the business tasks such as personnel management, financial management, and invoicing. The expert system module combines artificial intelligent technology such as data mining and machine learning to (1) realize egg shape index analysis; (2) provide feeding standards, breeding recommendations, mortality analysis, and other functions, and (3) realize poultry disease diagnosis and an early warning system based on audio and image analysis of poultry. Modules are functionally independent, with data-sharing capability.

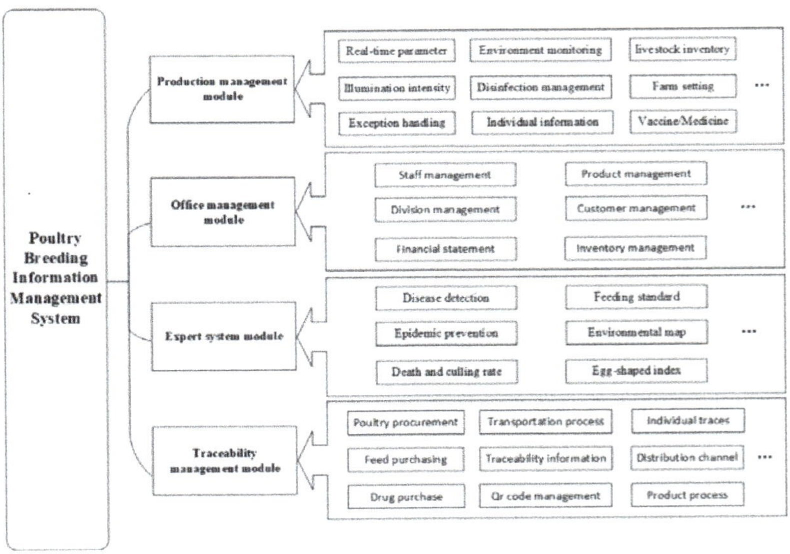

Figure 2. Functional framework of poultry farming information management system.

3. Wireless Sensor Network Design

The bottom layer of the system is mainly composed of wireless sensor networks, which are used to manage terminal nodes and the uploaded data information.

The composition of the wireless sensor network in the poultry house is shown in Figure 3. Each terminal node uses AT commands to automatically search for the wireless network by name and join it. After joining the network, it can independently obtain the device IP address and server IP address. Open the transparent transmission mode through the AT command, and use the UDP transmission protocol to transmit the data information. At the same time, in order to summarize and forward the data information uploaded by each terminal node, a data server should also be configured in the wireless sensor network in the poultry house.

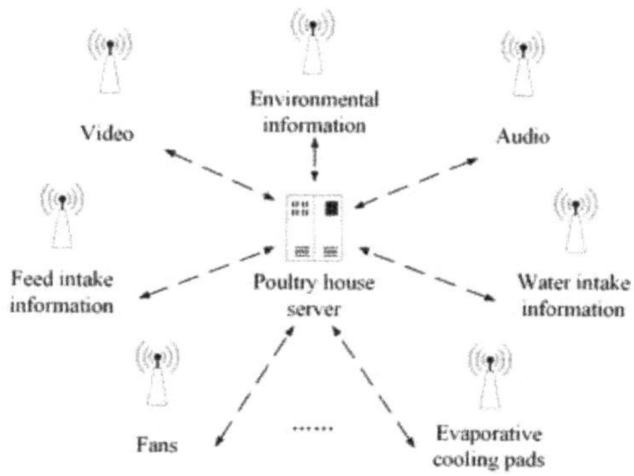

Figure 3. The composition of the wireless sensor network in the poultry house.

The wireless sensor network and various types of intelligent equipment can solve the problem of obtaining and transmitting various kinds of rearing information (e.g., environmental information, poultry weight information, poultry dietary information). This section mainly takes environmental information as an example to introduce the implementation process of information acquisition and upload. In the example, four environmental information sensing units are deployed in the wireless sensor network.

3.1. Poultry House Server

A data server should be configured in the wireless sensor network in the poultry house to process and forward the data information obtained by the terminal nodes in the network. Therefore, the server in the poultry house should choose a controller that has a processor, operating system, wireless network card, and can store programs. Considering the harsh working environment in the poultry house, this research selected the Industrial Personal Computer (IPC) with stronger waterproof, dustproof, and anti-interference capabilities than the data server.

In addition to the attributes and characteristics of ordinary computers, it also has stronger anti-interference ability and long-term uninterrupted working ability, which are suitable for use in the context of poultry farming environment. This research work selects an industrial control computer as the server in the poultry house to process and upload the data generated by the terminal node of the wireless sensor network, and the performance parameters are shown in Table 2.

Table 2. Industrial control computer performance parameters.

Device	Parameter	Manufacturer
CPU	Intel Core i5-7440HQ @ 2.80 GHz	Intel
RAM	8 GB (DDR4 2666 MHz)	SAMSUNG
Operating system	Windows 10 Professional 64-bit	Microsoft
Hard disk	NT-128 (128 GB/SSD)	Kingspec
Network card	43224AG 802.11 n Wi-Fi Adapter	Broadcom Corporation

In order to realize the processing and uploading of the data information generated by the terminal node, and at the same time realize the management and control of the wireless sensor network in the poultry house, a set of server programs is designed and loaded on the server in the poultry house to achieve the above-mentioned purpose.

3.2. Environmental Information Sensing Unit (EISU)

As the terminal node of the wireless sensor network, the environmental information sensing unit (EISU) in the poultry house should contain various digital environmental sensors, such as temperature and humidity sensors, carbon dioxide concentration sensors, hydrogen sulfide concentration sensors, light intensity sensors, wind speed sensors, etc., and it should also be equipped with a micro processor and wireless transmission module. The structure diagram is shown in Figure 4, and the performance parameters of the environmental sensors used are shown in Table 3, which are provided by the manufacturer.

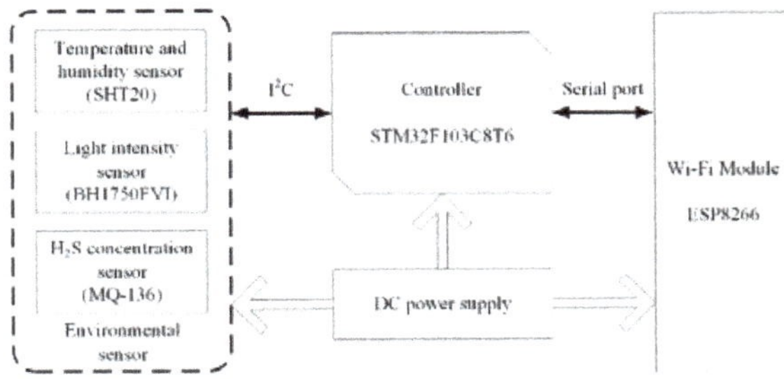

Figure 4. The structure diagram of the environmental information sensing unit structure.

Table 3. The environmental sensors performance parameters.

Type	Range	Resolution	Accuracy	Model
Temperature/°C	−40~125	0.01	±0.3 °C	SHT20
Relative humidity/%	0~100	0.01	±3%	
Light intensity/lx	0~65,535	0.01	±20%	BH1750FVI
H_2S concentration/ppm	1~200	0.1	±3%	MQ-136

Various environmental sensors are used to sense and measure the parameter values of the surrounding environment and using the I2C protocol to transmit them to the microcontroller through the data bus. The microcontroller is responsible for packaging the environmental data information according to the data packet format in Table 4 and uploading it to the poultry house server through the wireless transmission module; the diagram of the data flow is shown in Figure 5. Finally, the poultry house server uploads the data to the cloud database.

Table 4. Environmental information packet format.

Number	Identifier	Data (Hex)	Size (Byte)	Description
1	EI	45 49	2	Packet header
2	PL	-	4	Packet length
3	UN	55 4E	2	EISU number
4	TS	54 53	2	Temperature data start flag
5	TD	-	4	Temperature data
6	TE	54 45	2	Temperature data end flag
7	HS	48 53	2	Humidity data start flag
8	HD	-	4	Humidity data
9	HE	48 45	2	Humidity data end flag
10	BS	42 53	2	Light intensity data start flag
11	BD	-	4	Light intensity data
12	BE	42 45	2	Light intensity data end flag
13	SS	53 53	2	H_2S concentration data start flag
14	SD	-	4	H_2S concentration data
15	SE	53 45	2	H_2S concentration data end flag
16	CRC	-	4	Check code
17	EOP	FF 45	2	End of packet flag

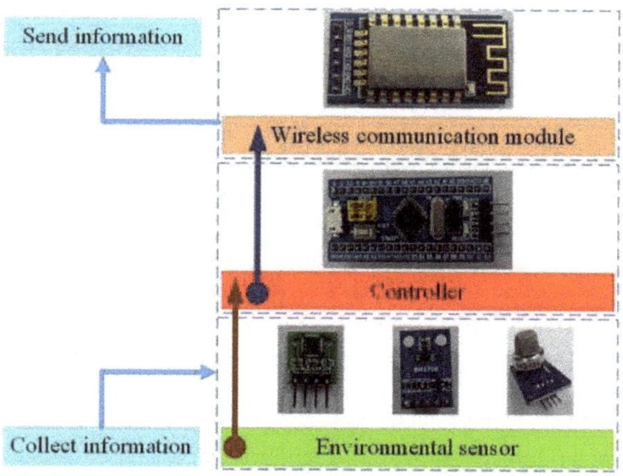

Figure 5. Flow chart of multi-threaded processing environment information data program.

3.3. Data Collection and Transmission

During the data transmission process of the wireless sensor network in the poultry house, there is no need to establish a one-to-one connection, only a one-to-many connection is needed to realize the communication between the terminal node and the server. Due to the large amount of data information transferred between the terminal node and the server and the number of transmissions, combined with the actual functional requirements, the network communication between the terminal node and the server of the wireless sensor network in the poultry house mainly uses UDP transmission protocol and socket technology.

The terminal node packages the data in a certain format and uses socket technology to send messages to the specific IP address and port number of the poultry house server. When the data information is successfully sent to the server in the poultry house, the server program downloads the data packet from the monitored port; then, it uses the specified transmission format to split the information, and it executes the corresponding SQL statement according to the type of data to upload the data to the cloud database; the data processing flow chart is shown in Figure 6.

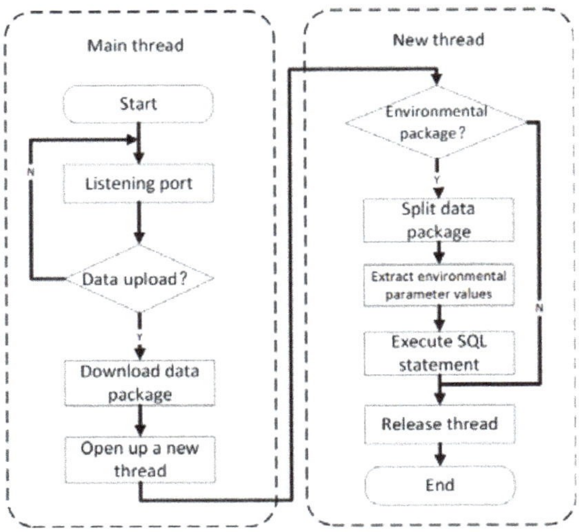

Figure 6. Data flow diagram of environmental information sensing unit.

Since there are many data transmission devices placed in the poultry house, it is necessary to set a routing node in the poultry house to forward the network data package to the cloud database. Taking into account the problem of network fluctuations, the server in the house may be disconnected from the cloud database, and the data information generated in the poultry house has real-time and continuous characteristics, so a data protection program must be designed to prevent data loss information. Figure 7 shows the local transmission strategy of the routing node, taking into account the network factors such that the system data security is improved.

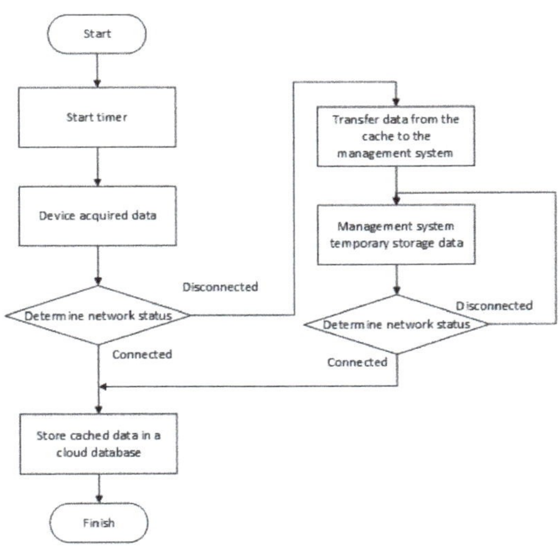

Figure 7. Local transmission strategy of the routing node flow chart.

4. Management System Implementation

The process of constructing the poultry farming information management system could be modularized, constructed, and tested one by one. First, a comprehensive platform for the management system can be built, and then the functional modules can be assembled. Figure 8 illustrates the home page of the management system.

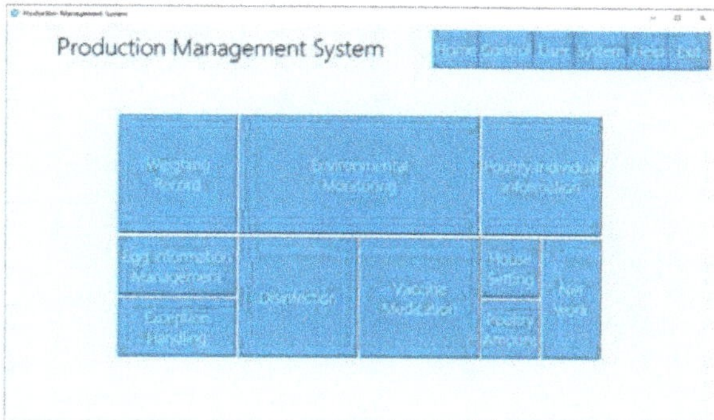

Figure 8. Management system home, includes ten function modules: weighing record, environmental monitoring, poultry individual information, egg information management, exception handing, disinfection, vaccine medication, house setting, poultry amount, and network.

This section mainly introduces the design and implementation effects of the environmental information management module and the poultry disease monitoring module.

4.1. Environmental Information Management Module

This module mainly realizes real-time monitoring of environmental information in the poultry house, and it can display the fluctuations of different environmental parameters. It can provide early warning in time when the environment in the house is abnormal and at the same time carry out good storage and management of historical environmental data.

The business process diagram of this module is shown in Figure 9. The EISUs upload the collected environmental information to the cloud database through the poultry house server. The cloud database is responsible for storing and managing the information. The environmental information management module makes active requests, queries the cloud database, and then display the feedback data through this module, so as to realize the management function of environmental information in the poultry house. The software interface effect of this module is shown in Figure 10.

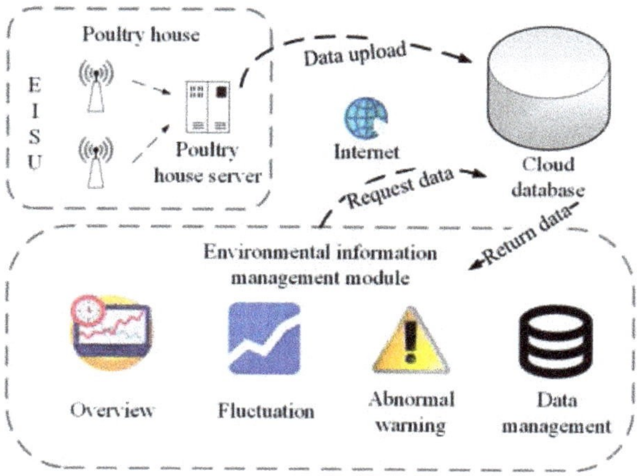

Figure 9. The business process chart of environmental information management module. EISU means environmental information sensing unit.

(**a**) (**b**)

Figure 10. The software interface effect of an environmental information management module, (**a**) fluctuating presentation of environmental data, (**b**) historical environmental data, including temperature, humidity, light, H_2S concentration, ammonia concentration, carbon dioxide concentration, etc.

4.2. Poultry Disease Monitoring Module

The disease monitoring and early warning function of this module is based on a large number of research results of our laboratory team, mainly through the analysis of poultry video and audio information to obtain information about the health status of poultry or related information about poultry disease [19–24].

The existing research foundation can be used to realize the detection function of poultry disease. On this basis, combined with the related work of this management system (poultry house environmental monitoring, poultry growth information monitoring, and diet and water consumption monitoring), it can further realize the function of monitoring the health status of poultry. The principle diagram of the method of monitoring the health status of poultry and early warning of disease is shown in Figure 11.

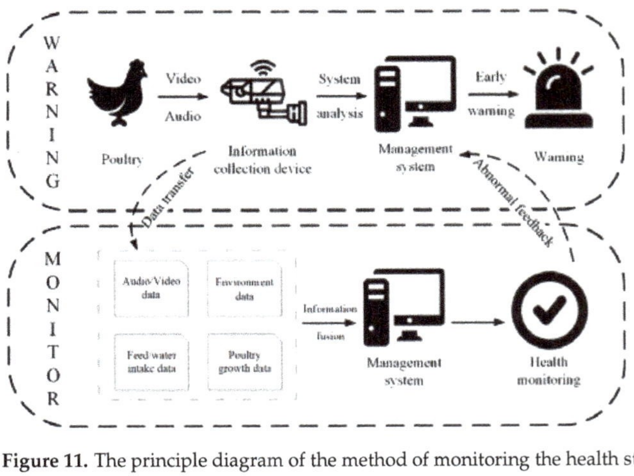

Figure 11. The principle diagram of the method of monitoring the health status of poultry and early warning of disease.

The software interface implemented by the disease detection function is shown in Figure 12. The poultry pictures and video content are used to monitor the health of the poultry. The detection principle and method are mainly to extract the features of the area to be detected under the food trough in the image, and they use the deep learning method to analyze and calculate the behavior of the chickens in the feature area and then judge the health of the chickens.

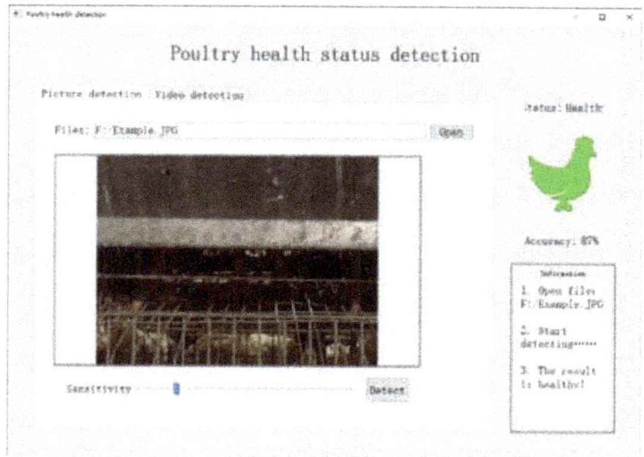

Figure 12. The software interface implemented by the disease detection function; by detecting the visual part of the poultry (the standing area below the feed trough), the current behavioral state of the poultry (standing, lying on the stomach) can be judged, and the health state of the chicken can be inferred.

5. System Performance Test

5.1. Wireless Sensor Network Testing

The system prototype constructed by this research is deployed in the breeder farm of South China Agricultural University. The test of the wireless sensor network system is mainly carried out on the farm. The on-site environment of the poultry house is shown in Figure 13. There are three rows of chicken cages, and each row is divided into three layers.

Figure 13. The poultry house of the signal strength and transmission rate test.

In order to analyze the performance of the constructed wireless sensor network in the poultry house, we tested the signal strength, transmission rate, and transmission stability of the environmental information sensing unit.

Figure 14 shows the test results of the signal strength and the transmission rate of the environmental information sensing unit. We tested the signal strength and transmission rate of the environmental information sensing unit under four conditions: (1) No external antennas, cages, or other obstacles; (2) No external antennas, but cages and other obstacles; (3) Equipped with an external antenna, which is not blocked by obstacles such as cages; and (4) Equipped with an external antenna, and is blocked by obstacles such as cages. Every condition was tested three times, and we took the average value as the signal strength and transmission rate after the signal stabilizes for about 30 s.

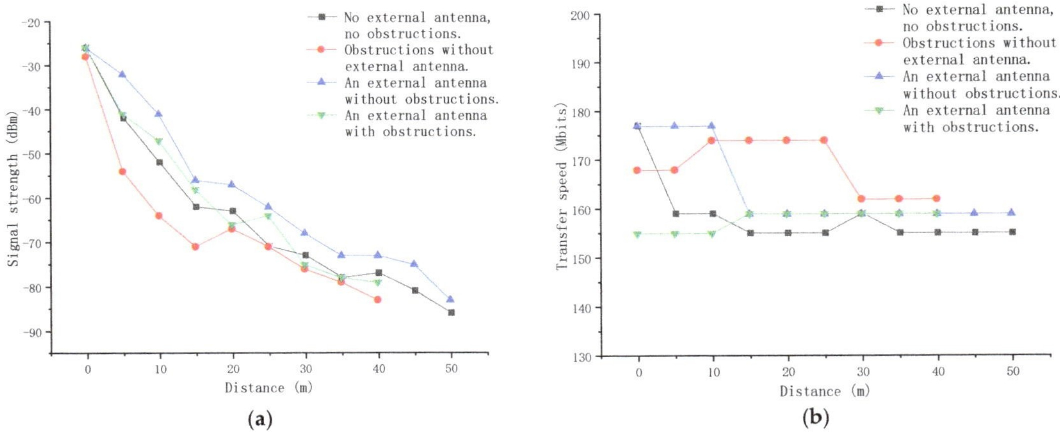

Figure 14. The test results of the environmental information sensing unit: (**a**) signal strength; (**b**) transmission rate.

The test results show that the signal strength of the environmental information sensing unit will decrease as the distance increases, and obstacles such as cages in the poultry house will also weakly affect the signal strength of the environmental information sensing unit; when using a 2 dBi gain antenna, it can effectively improve the signal strength of the environmental information sensing unit, but the transmission speed of the environmental information sensing unit is basically not affected by the distance, and its transmission speed has been stable at 150–180 Mbits. This shows that the environmental information sensing unit is equipped with a 2 dBi gain antenna, which can work normally in a poultry house with a radius of 40 m.

5.2. Cloud Database Testing

In this study, the cloud database system used Alibaba Cloud Service (RDS version of cloud database), the database version was MySQL 5.7, the storage engine used was InnoDB, the database memory was 1024 MB, the storage space was 20 GB, and the maximum number of supported connections was 2000. During the test, the performance of the cloud database was monitored for a period of time (one hour). During this period, there were four environmental information sensing units in the poultry house, two poultry house servers for continuous data uploading, and eight users who use the host computer software to read and write the database.

According to the test results of the cloud database performance parameters, the data query task demand is greater than the data upload task. However, the overall performance of the database is stable, the CPU and memory utilization rates are kept at a low level, and the cloud database system runs without pressure.

Figure 15 provides a more intuitive understanding of the operating status of the cloud database. During this period of time, the number of input/output operations per second (IOPS) of the cloud database was basically maintained at about 1.5, and the number of queries per second (QPS) remained above 13, but the utilization of the central processing unit (CPU) and memory has been maintained at a relatively low level (CPU utilization rate is 4.6–4.9%, memory utilization rate is 5.9–6.3%). It can be seen that the operation of the cloud database is stress-free under the experimental conditions, and the normal use of the system can be guaranteed.

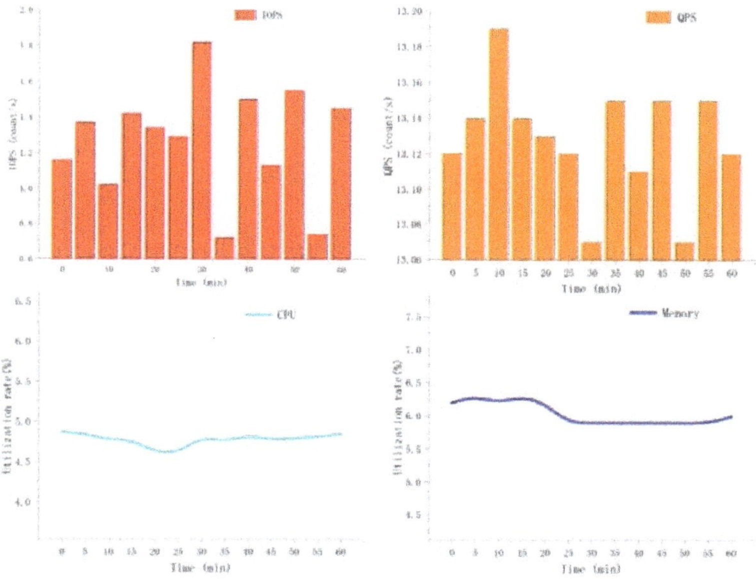

Figure 15. Cloud database performance parameters within one hour. IOPS means input and output operations per second, QPS means queries per second, and CPU means central processing unit.

6. Discussion

Designing a comprehensive and practical poultry farming information management system requires research on three aspects: (1) hardware design and networking of the underlying device, (2) database design and communication related issues, and (3) upper client software design.

LANs and servers should be deployed in poultry houses, and the use of wireless sensor networks (WSN) leads to low-cost and low-power deployments, making them the dominant choice [25]. The sensor is communicated as a child node in the local area network, and the data are connected to the external network through the server to upload the data to the cloud database for storage and query. The data format, transmission protocol, packet loss rate during transmission, and transmission strategy when network failure occurs should be further studied.

Environmental monitoring in poultry houses is the top priority to address animal welfare [26,27]. However, we have only studied a small part of the monitoring work of environmental parameters; there are still many important environmental parameters such as ammonia concentration, dust, and microorganisms monitoring work that need to be further studied. In addition, due to the uncertainty of hydrogen sulfide concentrations in the poultry house, we can consider using more sensitive hydrogen sulfide sensors [28].

How to effectively manage the massive information generated in the system is the primary goal of the future development [11,29]. Through good monitoring and proper storage of data, data mining technology can be used to diagnose poultry disease and provide early warning opportunities [30]. Moreover, using the cloud database as the core hub in this study could also help with the optimization of farming environment and breeding methods.

In order to meet the efficient management and the welfare of animal farming of a different scale poultry farm, this paper establishes a poultry farming information management system based on the cloud database, whose real-time monitoring of environmental information, poultry behavior information, and dietary information in poultry houses are integrated into the system. If successful, the system may meet the business needs of the poultry industry in regard to the environmental monitoring of poultry houses, monitoring of individual growth information, disease monitoring and early warning, traceability, and daily enterprise office management; at the same time, the data information generated during the production process will be well managed, and the poultry farming process informationized and intelligent [31].

7. Conclusions

The work reported in this paper builds a poultry farming information management system that covers the information management functions of production aquaculture, corporate office, product traceability, and poultry disease detection, and the unified construction of the system is realized. The system is divided into four modules according to the daily production management needs of farmers. By using intelligent sensors, building a wireless sensor network, and using a cloud database for data storage, a good interaction between modules, software, and hardware is realized, which can bring the animals closer to the farmer. At the same time, the system stores the collected data information in the cloud, and the cloud-based information management system will lead the development direction of the poultry farming management system.

Author Contributions: Conceptualization, H.Z. and T.Z.; methodology, T.Z. and C.F.; software, H.Z. and J.Z.; validation, H.Z., T.Z. and C.F.; data curation, H.Z. and J.Z.; writing—original draft preparation, H.Z.; writing—review and editing, T.Z. and C.F.; visualization, H.Z.; supervision, T.Z. and X.Y.; project administration, T.Z.; funding acquisition, T.Z. All authors have read and agreed to the published version of the manuscript.

Funding: This research was by National key research and development plan [grant No. 2018YFD0500705] and Guangdong Province Special Fund for Modern Agricultural Industry Common Key Technology R&D Innovation Team [grant No. 2020KJ129], China.

Institutional Review Board Statement: Not applicable.

Informed Consent Statement: Not applicable.

Data Availability Statement: The data presented in this study are available on request from the 1st Author (Haikun Zheng).

Acknowledgments: The authors appreciate the support and assistance provided by the staff of the poultry farm of South China Agricultural University.

Conflicts of Interest: The authors declare no conflict of interest.

References

1. Fang, C.; Huang, J.; Cuan, K.; Zhuang, X.; Zhang, T. Comparative study on poultry target tracking algorithms based on a deep regression network. *Biosyst. Eng.* **2020**, *190*, 176–183. [CrossRef]
2. Fang, C.; Zhang, T.; Zheng, H.; Huang, J.; Cuan, K. Pose estimation and behavior classification of broiler chickens based on deep neural networks. *Comput. Electron. Agric.* **2021**, *180*, 105863. [CrossRef]
3. Sun, Q.; Pan, H. RFID-Based Intelligent Management System of Poultry House. *Appl. Mech. Mater.* **2013**, *433*, 1511–1514. [CrossRef]
4. Chen, H.; Xin, H.; Teng, G.; Meng, C.; Du, X.; Mao, T.; Wang, C. Cloud-based data management system for automatic real-time data acquisition from large-scale laying-hen farms. *Int. J. Agric. Biol. Eng.* **2016**, *9*, 106–115.
5. Aydin, A.; Bahr, C.; Berckmans, D. A real-time monitoring tool to automatically measure the feed intakes of multiple broiler chickens by sound analysis. *Comput. Electron. Agric.* **2015**, *114*, 1–6. [CrossRef]
6. Choukidar, G.A.; Dawande, N.A. Smart Poultry Farm Automation and Monitoring System. In Proceedings of the 2017 International Conference on Computing, Communication, Control and Automation (ICCUBEA), Pune, India, 17–18 August 2017; pp. 1–5.
7. Pan, L.; Yang, S. A new intelligent electronic nose system for measuring and analysing livestock and poultry farm odours. *Environ. Monit. Assess.* **2007**, *135*, 399–408. [CrossRef] [PubMed]
8. Yu, L.; Teng, G.; Riskowski, G.L.; Xu, X.; Guo, W. Uncertainty analysis of a web-based data acquisition system for poultry management with sensor networks. *Eng. Agrícola* **2018**, *38*, 857–863. [CrossRef]
9. Irvine, R.M. A conceptual study of value chain analysis as a tool for assessing a veterinary surveillance system for poultry in Great Britain. *Agric. Syst.* **2015**, *135*, 143–158. [CrossRef]

10. So-In, C.; Poolsanguan, S.; Rujirakul, K. A hybrid mobile environmental and population density management system for smart poultry farms. *Comput. Electron. Agric.* **2014**, *109*, 287–301. [CrossRef]
11. Rowe, E.; Dawkins, M.S.; Gebhardt-Henrich, S.G. A Systematic Review of Precision Livestock Farming in the Poultry Sector: Is Technology Focussed on Improving Bird Welfare? *Animals* **2019**, *9*, 614. [CrossRef]
12. Norton, T.; Chen, C.; Larsen, M.L.V.; Berckmans, D. Review: Precision livestock farming: Building 'digital representations' to bring the animals closer to the farmer. *Animal* **2019**, *13*, 3009–3017. [CrossRef] [PubMed]
13. Balachandar, S.; Chinnaiyan, R. Internet of Things Based Reliable Real-Time Disease Monitoring of Poultry Farming Imagery Analytics. In Proceedings of the International Conference on Computer Networks, Big Data and IoT (ICCBI—2018), Madurai, India, 19–20 December 2018; Pandian, A., Senjyu, T., Islam, S., Wang, H., Eds.; Lecture Notes on Data Engineering and Communications Technologies. Springer: Cham, Switzerland, 2020; Volume 31. [CrossRef]
14. Stefanova, M. Precision Poultry Farming: Software Architecture Framework and Online Zootechnical Diary for Monitoring and Collaborating on Hens' Health. In *Information and Communication Technologies in Modern Agricultural Development*; Salampasis, M., Bournaris, T., Eds.; Springer: Cham, Switzerland, 2019. [CrossRef]
15. Astill, J.; Dara, R.A.; Fraser, E.D.G.; Roberts, B.; Sharif, S. Smart poultry management: Smart sensors, big data, and the internet of things. *Comput. Electron. Agric.* **2020**, *170*, 105291. [CrossRef]
16. Teng, G. Information sensing and environment control of precision facility livestock and poultry farming. *Smart Agric.* **2019**, *1*, 1–12.
17. Mirabella, O.; Brischetto, M. A hybrid wired/wireless networking infrastructure for greenhouse management. *IEEE Trans. Instrum. Meas.* **2011**, *60*, 398–407. [CrossRef]
18. Challoo, R.; Oladeinde, A.; Yilmazer, N.; Ozcelik, S.; Challoo, L. An overview and assessment of wireless technologies and co-existence of ZigBee, Bluetooth and Wi-Fi devices. *Procedia Comput. Sci.* **2012**, *12*, 386–391. [CrossRef]
19. Huang, J.; Wang, W.; Zhang, T. Method for detecting avian influenza disease of chickens based on sound analysis. *Biosyst. Eng.* **2019**, *180*, 16–24. [CrossRef]
20. Zhuang, X.; Bi, M.; Guo, J.; Wu, S.; Zhang, T. Development of an early warning algorithm to detect sick broilers. *Comput. Electron. Agric.* **2018**, *144*, 102–113. [CrossRef]
21. Zhuang, X.; Zhang, T. Detection of sick broilers by digital image processing and deep learning. *Biosyst. Eng.* **2019**, *179*, 106–116. [CrossRef]
22. Zhang, T.; Huang, J. Detection of chicken infected with avian influenza based on audio features and fuzzy neural network. *Trans. Chin. Soc. Agric. Eng.* **2019**, *35*, 168–174.
23. Huang, J.; Zhang, T.; Cuan, K.; Fang, C. An intelligent method for detecting poultry eating behaviour based on vocalization signals. *Comput. Electron. Agric.* **2021**, *180*, 105884. [CrossRef]
24. Cuan, K.; Zhang, T.; Huang, J.; Fang, C.; Guan, Y. Detection of avian influenza-infected chickens based on a chicken sound convolutional neural network. *Comput. Electron. Agric.* **2020**, *178*, 105688. [CrossRef]
25. Garcia-Sanchez, A.J.; Garcia-Sanchez, F.; Garcia-Haro, J. Wireless sensor network deployment for integrating video-surveillance and data-monitoring in precision agriculture over distributed crops. *Comput. Electron. Agric.* **2011**, *75*, 288–303. [CrossRef]
26. Gonzalez-Mora, A.F.; Larios, A.D.; Rousseau, A.N.; Godbout, S.; Morin, C.; Palacios, J.H.; Grenier, M.; Fournel, S. Assessing Environmental Control Strategies in Cage-Free Egg Production Systems: Effect on Spatial Occupancy and Natural Behaviors. *Animals* **2021**, *11*, 17. [CrossRef] [PubMed]
27. Adler, C.; Schmithausen, A.J.; Trimborn, M.; Heitmann, S.; Spindler, B.; Tiemann, I.; Kemper, N.; Büscher, W. Effects of a Partially Perforated Flooring System on Ammonia Emissions in Broiler Housing—Conflict of Objectives between Animal Welfare and Environment? *Animals* **2021**, *11*, 707. [CrossRef]
28. Lee, M.; Wi, J.; Koziel, J.A.; Ahn, H.; Li, P.; Chen, B.; Meiirkhanuly, Z.; Banik, C.; Jenks, W. Effects of UV-A Light Treatment on Ammonia, hydrogen sulfide, greenhouse gases, and ozone in simulated poultry barn conditions. *Atmosphere* **2020**, *11*, 283. [CrossRef]
29. Li, G.; Huang, Y.; Chen, Z.; Chesser, G.D., Jr.; Purswell, J.L.; Linhoss, J.; Zhao, Y. Practices and Applications of Convolutional Neural Network-Based Computer Vision Systems in Animal Farming: A Review. *Sensors* **2021**, *21*, 1492. [CrossRef] [PubMed]
30. Okinda, C.; Lu, M.; Liu, L.; Nyalala, I.; Muneri, C.; Wang, J.; Zhang, H.; Shen, M. A machine vision system for early detection and prediction of sick birds: A broiler chicken model. *Biosyst. Eng.* **2019**, *188*, 229–242. [CrossRef]
31. Sallabi, F.; Fadel, M.; Hussein, A.; Jaffar, A.; El Khatib, H. Design and implementation of an electronic mobile poultry production documentation system. *Comput. Electron. Agric.* **2011**, *76*, 28–37. [CrossRef]

animals

MDPI

Article

Characterizing Sounds of Different Sources in a Commercial Broiler House

Xiao Yang [1], Yang Zhao [1,*], Hairong Qi [2] and George T. Tabler [3]

[1] Department of Animal Science, The University of Tennessee, Knoxville, TN 37996, USA; xyang45@vols.utk.edu
[2] Department of Electrical and Computer Engineering, The University of Tennessee, Knoxville, TN 37996, USA; hqi@utk.edu
[3] Department of Poultry Science, Mississippi State University, Mississippi State, MS 39762, USA; ttabler@poultry.msstate.edu
* Correspondence: yzhao@utk.edu

Simple Summary: Acoustic signal in commercial broiler houses is a mixture of sounds from different sources. However, the characteristics of sounds from different sources have not been well understood. In this study, the sound frequency ranges of six common sounds, including bird vocalization, fan, feed system, heater, wing flapping and dustbathing, were determined; and their relations with bird age were investigated. The outcome of this research provides valuable information for using sound signal to monitor animal behavior and equipment operation.

Abstract: Audio data collected in commercial broiler houses are mixed sounds of different sources that contain useful information regarding bird health condition, bird behavior, and equipment operation. However, characterizations of the sounds of different sources in commercial broiler houses have not been well established. The objective of this study was, therefore, to determine the frequency ranges of six common sounds, including bird vocalization, fan, feed system, heater, wing flapping, and dustbathing, at bird ages of week 1 to 8 in a commercial Ross 708 broiler house. In addition, the frequencies of flapping (in wing flapping events, flaps/s) and scratching (during dustbathing, scratches/s) behaviors were examined through sound analysis. A microphone was installed in the middle of broiler house at the height of 40 cm above the back of birds to record audio data at a sampling frequency of 44,100 Hz. A top-view camera was installed to continuously monitor bird activities. Total of 85 min audio data were manually labeled and fed to MATLAB for analysis. The audio data were decomposed using Maximum Overlap Discrete Wavelet Transform (MODWT). Decompositions of the six concerned sound sources were then transformed with the Fast Fourier Transform (FFT) method to generate the single-sided amplitude spectrums. By fitting the amplitude spectrum of each sound source into a Gaussian regression model, its frequency range was determined as the span of the three standard deviations (99% CI) away from the mean. The behavioral frequencies were determined by examining the spectrograms of wing flapping and dustbathing sounds. They were calculated by dividing the number of movements by the time duration of complete behavioral events. The frequency ranges of bird vocalization changed from 2481 ± 191–4409 ± 136 Hz to 1058 ± 123–2501 ± 88 Hz as birds grew. For the sound of fan, the frequency range increased from 129 ± 36–1141 ± 50 Hz to 454 ± 86–1449 ± 75 Hz over the flock. The sound frequencies of feed system, heater, wing flapping and dustbathing varied from 0 Hz to over 18,000 Hz. The behavioral frequencies of wing flapping were continuously decreased from week 3 (17 ± 4 flaps/s) to week 8 (10 ± 1 flaps/s). For dustbathing, the behavioral frequencies decreased from 16 ± 2 scratches/s in week 3 to 11 ± 1 scratches/s in week 6. In conclusion, characterizing sounds of different sound sources in commercial broiler houses provides useful information for further advanced acoustic analysis that may assist farm management in continuous monitoring of animal health and behavior. It should be noted that this study was conducted with one flock in a commercial house. The generalization of the results remains to be explored.

Citation: Yang, X.; Zhao, Y.; Qi, H.; Tabler, G.T. Characterizing Sounds of Different Sources in a Commercial Broiler House. *Animals* **2021**, *11*, 916. https://doi.org/10.3390/ani11030916

Academic Editor: Janice Siegford

Received: 21 February 2021
Accepted: 19 March 2021
Published: 23 March 2021

Keywords: poultry; acoustic; audio; frequency; behavior

1. Introduction

Sound is defined as vibrations that travel through the air or another medium in the form of waves. The sound in commercial broiler houses is a mixture originated from multiple sources and can be generally categorized into two groups: (1) animal-based sounds and (2) equipment-based sounds. Animal-based sounds can be further categorized into animal vocalization (produced by the vibration of bird vocal cord) and behavioral sounds (produced during bird behaviors such as wing flapping and scratching). Equipment-based sounds refer to those produced during feed system, fans-on and heaters-on period.

Understanding the characteristics of sounds from different sources in broiler houses allows further sound analysis that may assist farmers in farm management and welfare monitoring. Frequency is one of the most important characteristics of sound. Audio signal in time domain only reflects the loudness, therefore, most of audio signal processing techniques and algorithms involve frequency analysis using techniques such as Fourier Transform [1], filtering [2], spectrogram [3], etc., for source identification.

In recent years, sound analysis as a non-invasive method has become an increasingly important tool in animal disease detection, behavior monitoring and welfare determination [4–6]. Cuan et al. [7] proposed a sound recognition method based on convolutional neural network to detect the infection of avian influenza, yielding 90% accuracy. Chung et al. [8] found that audio data can accurately detect (94% accuracy) and recognize (91% accuracy) pig wasting diseases. A sound-based product (SoundTalks NV., Leuven, Belgium) was commercialized to continuously and automatically detect pig respiratory disease at an early stage [9]. For animal behaviors, several studies were conducted to identify the feeding behavior of broiler chickens by analyzing audio data [10,11]. It has also been shown that peak frequency of bird vocalization can serve as the indicator of broiler age [12] and body weight [13].

In previous studies, the equipment-based sounds that are produced by mechanical systems were mostly considered as noise. The negative effects of farm noise on animal welfare have been reported [14–16]. However, as unavoidable acoustic sources in conventional broiler houses, equipment-based sounds also contain important information that can be used for farm management. For instance, the sound of the feed system can be an indicator for proper operation of the feed system, and the sound of fans for proper ventilation. On the other hand, understanding the characteristics of the equipment-based sounds may better help to remove these background sounds when only animal-based sounds are the concerns.

The objective of this study was to determine the frequency ranges of six common sounds, including bird vocalization, fan, feed system, heater, wing flapping and dustbathing in a commercial broiler house over an eight-week production cycle. In addition, the frequencies of flapping (in wing flapping events, flaps/sec) and scratching (during dustbathing, scratches/sec) behaviors were examined through sound analysis.

2. Materials and Methods

2.1. Housing, Animals and Management

The study was conducted in a commercial broiler house (east–west) located at Mississippi State University during May–June 2020. The house measured 120 × 13 × 3 m (L × W × H) with a capacity of 13,700 Ross 708 straight run broilers and a production cycle of eight weeks. The average slaughter body weight was 4.25 kg. All chicks were provided by a contract integrator in Mississippi. Flock management and diets followed the typical procedures in the industry. The lighting schedule was set to 24 L:0 D from 1 d to 7 d, 20 L:4 D from 8 d to 56 d. The light intensity was set to 54 lux from 1 d to 13 d, then gradually dimmed to 3 lux by 20 d and kept at 3 lux till 56 d. Lights were turned on at 05:00 h and turned off at 01:00 h of the next day. A total of 10 consistent speed fans (Acme BDR48J-A, 48", Acme Engineering & Manufacturing Corp., Muskogee, OK, USA) were installed in the broiler house, with six fans across the east end wall, two fans on the north side wall, and two fans on the south side wall. The air flow rate was 34,660 m^3/h at the static pressure of 0.05. The number and running time of fans were controlled by the house controller based on indoor air house temperature. As the experiment was conducted in summer, all fans were used by the end of the flock. A total of 13 heaters (Hog Slat GRO40, direct spark ignition, natural gas, 11.7 kW, Hog Slat Inc., Newton Grove, NC, USA) were distributed across the house and were installed 1.8 m above the floor. Heater on/off was controlled by the house controller as well. Automatic feeding system (Chore-Time, Revolution, A Division of CTB, Inc., Milford, IN, USA) was installed in the house. Feed was delivered along the feed line by an open coil auger inside the feed line tube and dropped into feeder pans halfway.

2.2. Audio Data Collection and Camera System

An H2n handy recorder (Zoom North American Inc., Hauppauge, NY, USA) (Figure 1a) was installed in the middle of broiler house at the height of 40 cm above the back of birds, as shown in Figure 1b. The recorder was 67.6 × 113.85 × 42.7 mm (W × H × D) in size and 130 g in weight. It was powered by two AA batteries or by an AD-17 USB to AC adapter (used in this study). Up to 120 dB sound can be captured. The sampling frequency was set to 44,100 Hz in this study. Audio signal was continuously recorded and saved to a 2 GB micro SD card and exported weekly.

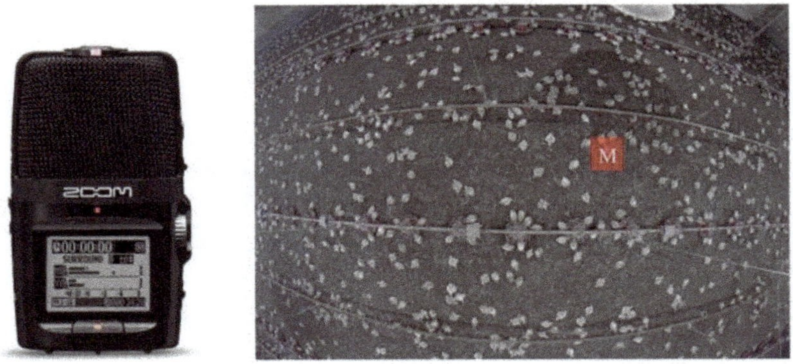

(a) (b)

Figure 1. Image of the recorder (**a**) and location of the recorder M (**b**).

A fisheye IP camera (Dahua, IPC-EW4431-ASW, Dahua Technology USA Inc., Irvine, CA, USA) was installed on the ceiling (height = 3 m) right above the microphone to monitor the wing flapping and dustbathing behaviors of broilers. The frame rate of the camera was 25 frames per second.

2.3. Sound Discription

Six specific types of sound signals, including bird vocalization, fan, feed system, heater, wing flapping and dustbathing, were examined in this study. The sound of the fans was produced when mixing fans and/or ventilation fans were working. As the microphone was installed around 50 m away from the fan, the sound of the fan was identified as the sound of wind from the recorded audios. The sound of the feed system was produced during the period when feeder augers were running for feed delivery. The sound of the heaters was produced during the first week when heaters were operating. Wing flapping was identified as a bout of continuous, rapid flapping behaviors [17]. Dustbathing was defined as birds performing classic vertical wing shakes, and performing side-rubs or prone leg scratches [18]. In this study, dustbathing specifically refers to the behavior of scratching the litter.

2.4. Audio Signal Labeling and Pre-Processing

The software Audacity (v.2.4.2, Audacity Team, GNU General Public License (GPL)) was used to visualize and identify the signal by comparing with the recorded videos. A summary of labeled data is shown in Table 1. Bird vocalization, fan, feed system and heater were trimmed into multiple 10-s clips. As the wing flapping and dustbathing usually lasted for a short time, behavioral data were trimmed into multiple 1-s clips. Audio clips were then fed into MATLAB (2018b, The MathWorks, Inc., Natick, MA, USA) for further analysis. Spectrograms were used for an initial visual check on frequency ranges, as shown in Figure 2. The frequencies of bird vocalization (Figure 2a) and fan (Figure 2b) ranged 0–5000 Hz. Sounds frequencies of the feed system (Figure 2c), heater (Figure 2d), wing flapping (Figure 2e) and dustbathing (Figure 2f) ranged 0–19,000 Hz.

Table 1. Number of events and equivalent time labeled for each type of sound over eight weeks.

Sound Type	Total Time Labeled (min)	Total Number of Clips
Bird vocalization	40.0	240
Fan	40.0	240
Feed system	40.0	240
Heater	5.0	30
Wing flapping	6.7	402
Dustbathing	4.1	246

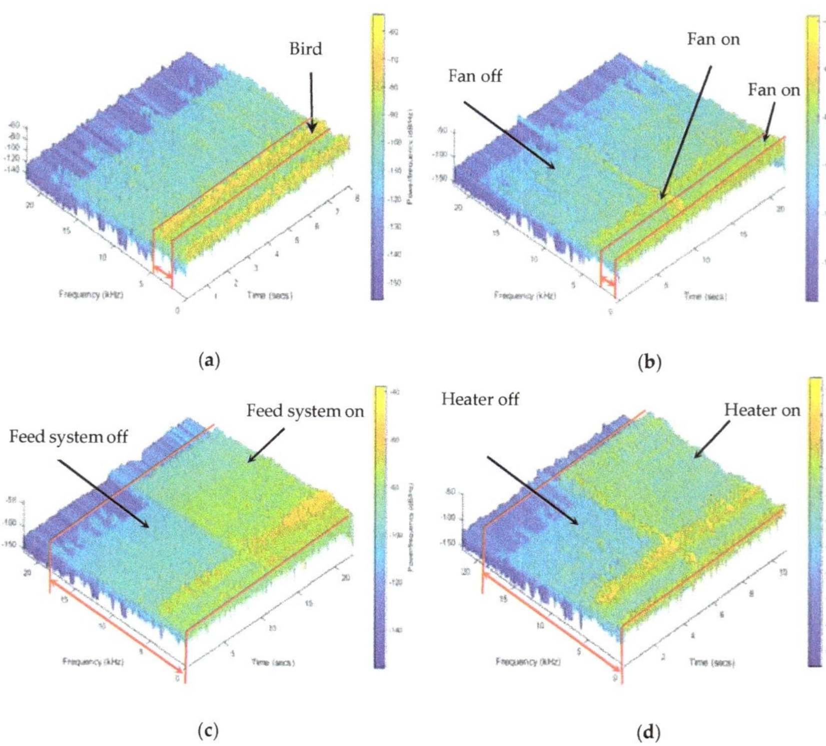

Figure 2. *Cont.*

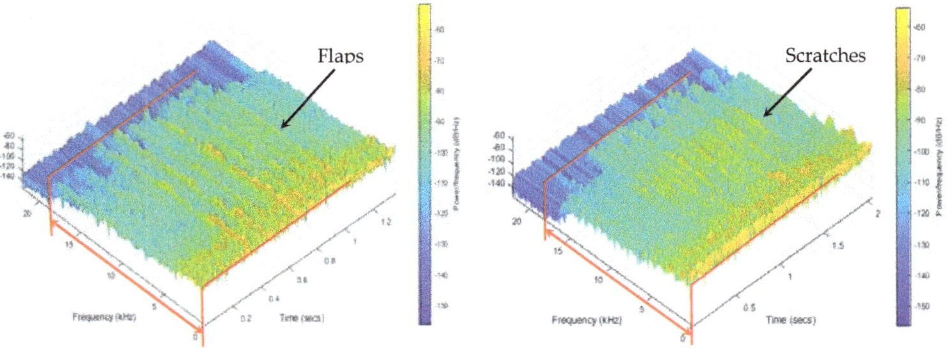

(e) (f)

Figure 2. Example signals of bird vocalization (**a**), fan (**b**), feed system (**c**), heater (**d**), wing flapping (**e**) and dustbathing (**f**) in frequency domain.

2.5. Maximum Overlap Discrete Wavelet Transform (MODWT)

The MODWT is a linear filtering operation that transforms a signal into multilevel wavelet and scaling coefficients [19]. As the flowchart for three-level MODWT shown in Figure 3, the MODWT applies low/high-pass filters to split the frequency components of the input signal into different scales. The filters are determined depending on the mother wavelet, which is selected in advance. The mother wavelets include Daubechies wavelet, Coiflects wavelet, Haar wavelet, and Symlet wavelet, and so on. Details on different wavelets can be found in publications by Percival and Walden [20]. The default mother wavelet (Symlet, N = 4) in MATLAB was used in our study. The decomposition level was determined based on the Equation (1) [21].

$$L = int[log(n)] \tag{1}$$

where *int*[] is the function that returns the nearest integer of a number and n is the data length. For the study, n = 441,000, L = 6. The decompositions of bird vocalization signals are shown in Figure 4.

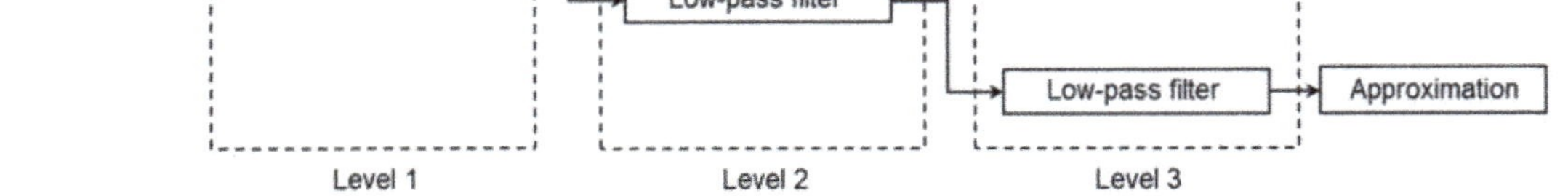

Figure 3. Flowchart for three-level Maximum Overlap Discrete Wavelet Transform.

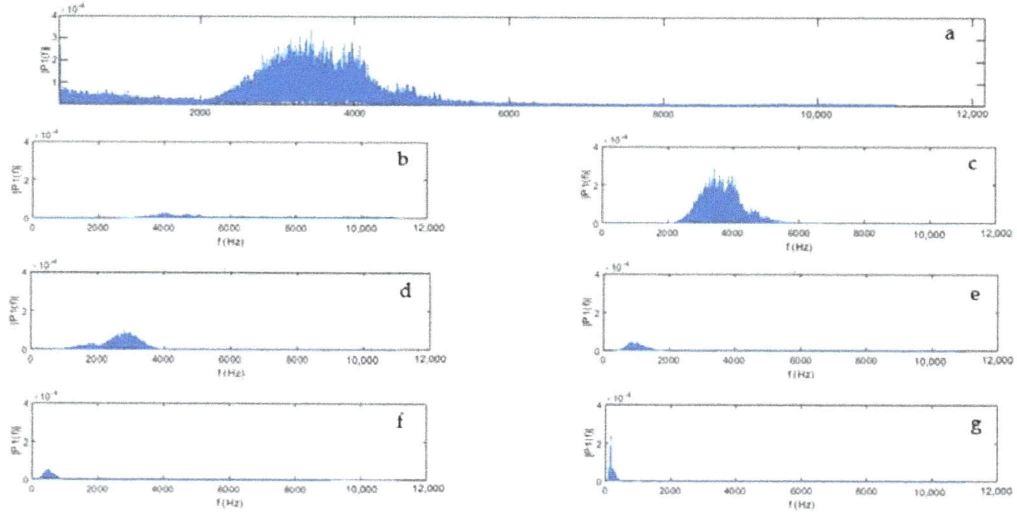

Figure 4. Single-sided amplitude spectrum of original signal (**a**) and decompositions 1–6 (**b**–**g**). Decomposition 2 (**c**) refers to bird vocalization, decomposition 4 (**e**) refers to fan.

2.6. Signal Processing

Two sounds with narrow frequency ranges (bird vocalization and fan) were decomposed using MODWT. Decompositions of both sounds were then transformed with the Fast Fourier Transform (FFT) method to generate the single-sided amplitude spectrums. By fitting the amplitude spectrum of each sound source into a Gaussian regression model, its frequency range was determined as the span of the three standard deviations (99% CI) away from the mean (Figure 5a). For those sounds with wide frequency ranges (feed system, heater, wing flapping and dustbathing), the upper ranges were determined by reading the largest frequency value from the FFT plot. The behavioral frequencies were determined by examining the occurrence of peak amplitudes in spectrograms during wing flapping and dustbathing (Figure 5b). They were calculated by dividing the number of wing flaps and scratches by the time duration of complete behavioral events.

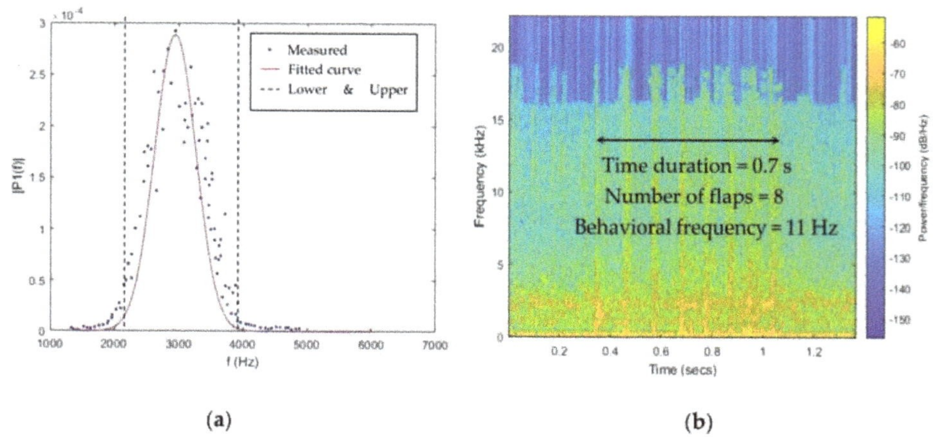

Figure 5. Example of Gaussian regression (**a**) and calculation of behavioral frequency (**b**). The sound sources of two sounds are bird vocalization (**a**) and wing flapping (**b**), respectively.

2.7. Statistic Analysis

The effects of bird age on each different sound frequency range were analyzed using the PROC GLM (generalized linear model) procedure in SAS 10.9 (SAS Institute., Cary, NC, USA). The differences in behavioral frequency between wing flapping and dustbathing were also tested. A significant difference in multiple comparisons of group means was defined as $p < 0.05$.

3. Results

3.1. Bird Vocalization

Table 2 shows the lower and upper frequency limits of bird vocalization at different ages. Over the flock, the highest bird vocalization frequency was 4409 Hz, and the lowest was 1058 Hz. In general, both lower and upper limits continuously decreased as the bird grew, but the decrease was not linear. Frequencies dropped faster in the first few weeks. From week 1 to 4, the lower and upper limits decreased 1020 Hz and 1665 Hz, respectively. From week 4 to 8, the lower and upper limits decreased 403 Hz and 292 Hz, respectively. The frequency ranges during the first three weeks were larger than week 4 to 8.

Table 2. Frequency range of bird vocalization at different bird ages (mean ± SD).

Bird Age (Week)	Lower Limit (Hz)	Upper Limit (Hz)
1	2481 ± 191^{a}	4409 ± 136^{a}
2	2038 ± 201^{b}	4289 ± 89^{b}
3	1889 ± 307^{c}	3997 ± 128^{c}
4	1461 ± 187^{d}	2744 ± 155^{d}
5	1418 ± 164^{d}	2668 ± 114^{e}
6	1190 ± 154^{e}	2628 ± 113^{e}
7	1100 ± 148^{ef}	2615 ± 118^{e}
8	1058 ± 123^{f}	2501 ± 88^{f}
average	1579	3231

^{a,b,c,d,e,f} Means in the same column with different superscripts are different ($p < 0.05$).

3.2. Fan

Table 3 shows the upper and lower frequency limits of the fan at different bird ages. Generally, the lower and upper frequency limits of the fan increased as the bird grew. From week 1 to 8, the highest sound frequency of the fan was 1203 Hz, and the lowest was 305 Hz. In weeks 5–8, the lower frequency limits were significantly higher than other weeks. No significant difference was found in upper limits in weeks 2–4. The upper limits of weeks 6–8 were significantly higher than other weeks. The frequency range varied between 716 Hz (week 1) and 791 Hz (week 3).

Table 3. Frequency range of fan at different bird ages (mean ± SD).

Bird Age (Week)	Lower Limit (Hz)	Upper Limit (Hz)
1	353 ± 62^{c}	1069 ± 40^{d}
2	331 ± 38^{cd}	1080 ± 46^{cd}
3	312 ± 40^{d}	1103 ± 52^{c}
4	335 ± 32^{cd}	1101 ± 21^{c}
5	407 ± 81^{b}	1161 ± 78^{b}
6	446 ± 28^{a}	1191 ± 27^{a}
7	417 ± 34^{b}	1203 ± 36^{a}
8	428 ± 44^{a}	1200 ± 36^{a}
average	379	1139

^{a,b,c,d} Means in the same column with different superscripts are different ($p < 0.05$).

3.3. Upper Limits of Feed System, Heater, Wing Flapping and Dustbathing

Table 4 shows the upper frequency limits of feed system, heater, wing flapping and dustbathing at different bird ages. In general, the frequency ranges of all the above sounds spanned widely from 0 to 19,000 Hz. For the feed system, the upper limits in the first two weeks were significantly lower than other weeks. In addition, the upper limit in week 8 was significantly higher than other weeks. The upper limit of the heater was the highest among the four types of sound. For wing flapping, the highest upper limit was observed in week 8, and the lowest was found in week 5. For dustbathing, the upper limit in week 3 was significantly lower than weeks 4–6. No significant difference was observed among weeks 4–6. When comparing two behavioral sounds, the average upper limit of wing flapping was higher than that of dustbathing.

Table 4. Upper frequency limits of feed system, heater, wing flapping and dustbathing at different bird ages (mean ± SD).

Bird Age (Week)	Feed System (Hz)	Heater (Hz)	Wing Flapping (Hz)	Dustbathing (Hz)
1	18,694 ± 149 ^{c}	18903 ± 24	-	-
2	18,655 ± 244 ^{c}	-	-	-
3	18,781 ± 61 ^{b}	-	18,830 ± 27 ^{ab}	18,771 ± 26 ^{b}
4	18,819 ± 45 ^{ab}	-	18,833 ± 50 ^{a}	18,793 ± 27 ^{a}
5	18,804 ± 54 ^{ab}	-	18,819± 27 ^{b}	18,791 ± 30 ^{a}
6	18,813 ± 113 ^{ab}	-	18,832± 34 ^{ab}	18,797 ± 32 ^{a}
7	18,833 ± 40 ^{ab}	-	18,829 ± 27 ^{ab}	-
8	18,857 ± 25 ^{a}	-	18,837 ± 28 ^{a}	-
average	18,782	18,903	18,830	18,788

^{a,b,c} Means in the same column with different superscripts are different ($p < 0.05$). Heater was only operated in week 1. The sound of wing flapping and dustbathing can be detected by microphone after week 2. No dustbathing behavior was identified in weeks 7 and 8.

3.4. Behavioral Frequency

Table 5 shows the behavioral frequencies of wing flapping and dustbathing at different bird ages. The behavioral frequencies of wing flapping and dustbathing were continuously decreased as the bird grew. In week 3, the behavioral frequencies of both wing flapping and dustbathing were significantly higher than other weeks. No significant difference was observed among weeks 4–6 for wing flapping and dustbathing. The behavioral frequency of wing flapping was significantly higher than that of dustbathing in weeks 4–6 ($p = 0.049$, $p = 0.0003$ and $p = 0.0001$, respectively).

Table 5. Behavioral frequency of wing flapping and dustbathing at different bird ages (mean ± SD).

Bird Age (Week)	Wing Flapping (Hz)	Dustbathing (Hz)
1	-	-
2	-	-
3	17 ± 4 ^{Aa}	16 ± 2 ^{Aa}
4	14 ± 3 ^{Ab}	12 ± 1 ^{Bb}
5	14 ± 3 ^{Ab}	12 ± 2 ^{Bb}
6	13 ± 2 ^{Ab}	11 ± 1 ^{Bb}
7	11 ± 2 ^{c}	-
8	10 ± 1 ^{c}	-

^{a,b,c} Means in the same column with different superscripts are different ($p < 0.05$). ^{A,B} Means in the same row with different superscripts are different ($p < 0.05$). The sound of wing flapping and dustbathing can be detected by microphone after week 3. No dustbathing was found on week 8.

4. Discussion

The frequency range of bird vocalization continuously decreased as the birds got older (Table 2). The result is consistent with that previously reported by Fontana et al. [22], who found a negative relationship between broiler vocalization frequency and bird age. The key assumption underlying the result is that larger animals often produce vocalizations with lower frequency than smaller animals. Vocalizations can be simply described as the result of tissue vibrations generated by the passage of air through a constriction in an animal's vocal tract [23]. Due to the physical and energetic constrains, animals cannot efficiently produce sound waves larger than the size of their body or their sound-producing apparatus [24]. According to the theory, bird vocalization has been used to automatically and continuously monitor broiler body weight [13].

Different frequency-based filters have been adopted to remove the ventilation noise at pre-processing stage [12,25,26]. However, the frequency of sound produced by fan operation in commercial broiler houses remains to be understood. In this study, we found that the upper frequency range of the fan in commercial broiler houses varied between 1069–1203 Hz, which is slightly higher than those reported (1000 Hz) in previous studies [13,25]. Furthermore, our results show that both lower and upper limits of fan sound frequency generally increased as the birds grew, probably due to the increased ventilation rate and air speed. As birds got older and the weather got hotter, more fans were required for higher ventilation rates and air speed to exhaust excess heat production by birds. If the size of air inlets does not change accordingly to maintain a proper static pressure (e.g., 25 Pa), the airflow going through each fan will alter, which may affect the air interaction with fan blades and the fan sound frequency.

The frequency of feed system and heater ranged from 0 to 19,000 Hz. The result indicates that the sound produced during feed system and heater operation cannot be simply removed by adding a bandpass filter. Other noise reduction methods will be needed. There was statistical difference in the upper limits of the automatic feeder sounds among weeks, which was possibly due to the differences in feed particle sizes that affected the frictions between augers and feed tubes during feed delivery. However, the changes of the upper limits were small.

Acoustic signal can be a useful tool for learning animal behaviors. Microphones have been widely used to detect the animal behaviors, e.g., foraging of beef cattle [27,28], chewing of dairy cows [29], and feeding behavior of broilers [10]. However, using audio signal to determine broiler wing flapping and dustbathing behaviors has not been reported before. Our results showed the upper frequency range limits of these two behavioral sounds varied between 18,770–18840 Hz. Therefore, in order to avoid information loss of the signal, adding a filter < 19,000 Hz is not recommended before analyzing wing flapping and dustbathing behaviors at pre-processing. From the spectrograms, both wing flapping and dustbathing showed unique patterns in time series, which may provide valuable information for behavior classification and recognition in future.

During the first two weeks, wing flapping and dustbathing behaviors can be observed in recorded videos; however, they were not efficiently captured by the microphone. This indicates the limitation of using a microphone to learn the behaviors of young chicks. No dustbathing was observed in weeks 7–8. Similar results were reported by Meluzzi et al. [30], that dustbathing activity was decreased as broilers get older. Litter quality could be one of the factors. Broilers prefer to bathe at the area with loose and dry materials [31], while the litter is stiffer and moister at the end of flock. The behavioral frequency of wing flapping and dustbathing continuously decreased as birds grew. The possible reason is that older birds were more physically challenged to perform these behaviors due to the body weight. This hypothesis also indicates that the behavioral frequency could be another indicator of bird age. As no study has been conducted on broiler behavioral frequency so far, further investigation will be needed for the hypothesis.

As two types of non-invasive methods in animal research, cameras and microphones were often used separately. Both of them have proved to be efficiently deployed in precision livestock farming. It would be exciting if the two methods can be combined. Most off-the-shelf cameras come with both channels of video and audio, which can provide the system's eyes and ears. Endless possibilities could be achieved by adding a computer as the brain, which will eventually achieve smart broiler farming.

5. Conclusions

In this study, the sound frequency ranges of bird vocalization, fan, feed system, heater, wing flapping and dustbathing in a commercial Ross 708 broiler house at different bird ages were determined using signal processing. The behavioral frequencies of wing flapping and dustbathing were examined as well. We concluded that the frequency range of bird vocalization continuously decreased as birds grew. The sound frequency of the fan generally increased from week 1 to week 8. The upper frequency range of the feed system, heater, wing flapping and dustbathing exceeded 18,000 Hz. Significant negative correlation of age on behavioral frequencies of wing flapping and dustbathing were observed. In summary, both broiler vocalization and equipment-based sounds showed temporal variations. These findings provide important insights into broiler welfare, health, and behavior determination using signal processing. Generalizing the results to other housing systems and broiler breeds will require more data.

Author Contributions: Conceptualization, X.Y. and Y.Z.; data curation, X.Y.; formal analysis, X.Y.; funding acquisition, Y.Z.; investigation, X.Y., Y.Z. and H.Q.; project administration, Y.Z. and G.T.T.; methodology, X.Y., Y.Z. and H.Q.; supervision, Y.Z.; resources, Y.Z. and G.T.T.; writing—original draft, X.Y.; writing—review and editing, Y.Z., H.Q. and G.T.T. All authors have read and agreed to the published version of the manuscript.

Funding: This study was financially supported by the Foundation for Feed and Agriculture Research (FFAR) SMART Broiler Initiative and the Mississippi Agricultural and Forestry Experiment Station (MAFES) Special Research Initiative.

Institutional Review Board Statement: Not applicable.

Data Availability Statement: The data presented in this study are available on request from the corresponding author. The data are not publicly available due to restrictions by the collaborator broiler producer.

Acknowledgments: The authors appreciate the support and assistance provided by the farm staff at Mississippi State University.

Conflicts of Interest: The authors declare no conflict of interest. The funders had no role in the design of the study; in the collection, analyses, or interpretation of data; in the writing of the manuscript; or in the decision to publish the results.

References

1. Bracewell, R.N. *The Fourier Transform and Its Applications*; McGraw-Hill: New York, NY, USA, 1986; Volume 31999.
2. Christiano, L.J.; Fitzgerald, T.J. The band pass filter. *Int. Econ. Rev.* **2003**, *44*, 435–465. [CrossRef]
3. Dörfler, M.; Bammer, R.; Grill, T. Inside the spectrogram: Convolutional Neural Networks in audio processing. In Proceedings of the 2017 International Conference on Sampling Theory and Applications (SampTA), Tallinn, Estonia, 3–7 July 2017.
4. Deniz, N.N.; Chelotti, J.O.; Galli, J.R.; Planisich, A.M.; Larripa, M.J.; Rufiner, H.L.; Giovanini, L.L. Embedded system for real-time monitoring of foraging behavior of grazing cattle using acoustic signals. *Comput. Electron. Agric.* **2017**, *138*, 167–174. [CrossRef]
5. Sheng, H.; Zhang, S.; Zuo, L.; Duan, G.; Zhang, H.; Okinda, C.; Shen, M.; Chen, K.; Lu, M.; Norton, T. Construction of sheep forage intake estimation models based on sound analysis. *Biosyst. Eng.* **2020**, *192*, 144–158. [CrossRef]
6. Chelotti, J.O.; Vanrell, S.R.; Galli, J.R.; Giovanini, L.L.; Rufiner, H.L. A pattern recognition approach for detecting and classifying jaw movements in grazing cattle. *Comput. Electron. Agric.* **2018**, *145*, 83–91. [CrossRef]
7. Cuan, K.; Zhang, T.; Huang, J.; Fang, C.; Guan, Y. Detection of avian influenza-infected chickens based on a chicken sound convolutional neural network. *Comput. Electron. Agric.* **2020**, *178*, 105688. [CrossRef]
8. Chung, Y.; Oh, S.; Lee, J.; Park, D.; Chang, H.-H.; Kim, S. Automatic Detection and Recognition of Pig Wasting Diseases Using Sound Data in Audio Surveillance Systems. *Sensors* **2013**, *13*, 12929–12942. [CrossRef] [PubMed]
9. Hemeryck, M. The Pig Cough Monitor in the EU-PLF project results and multimodal data analysis in two case studies. In Proceedings of the 7th European Conference on Precision Livestock Farming (EC-PLF), Milan, Italy, 15–18 September 2015.
10. Aydin, A.; Berckmans, D. Using sound technology to automatically detect the short-term feeding behaviours of broiler chickens. *Comput. Electron. Agric.* **2016**, *121*, 25–31. [CrossRef]
11. Aydin, A.; Bähr, C.; Berckmans, D. A real-time monitoring tool to automatically measure the feed intakes of multiple broiler chickens by sound analysis. *Comput. Electron. Agric.* **2015**, *114*, 1–6. [CrossRef]
12. Fontana, I.; Tullo, E.; Scrase, A.; Butterworth, A. Vocalisation sound pattern identification in young broiler chickens. *Animal* **2016**, *10*, 1567–1574. [CrossRef] [PubMed]

13. Fontana, I.; Tullo, E.; Carpentier, L.; Berckmans, D.; Butterworth, A.; Vranken, E.; Norton, T.; Berckmans, D.; Guarino, M. Sound analysis to model weight of broiler chickens. *Poult. Sci.* **2017**, *96*, 3938–3943. [CrossRef]
14. Otten, W.; Kanitz, E.; Puppe, B.; Tuchscherer, M.; Brüssow, K.; Nürnberg, G.; Stabenow, B. Acute and long term effects of chronic intermittent noise stress on hypothalamic-pituitary-adrenocortical and sympatho-adrenomedullary axis in pigs. *Anim. Sci.* **2004**, *78*, 271–283. [CrossRef]
15. Rabin, L.A. Anthropogenic noise and its effect on animal communication: An interface between comparative psychology and conservation biology. *Int. J. Comp. Psychol.* **2003**, *16*, 172–192.
16. Campo, J.; Gil, M.; Dávila, S. Effects of specific noise and music stimuli on stress and fear levels of laying hens of several breeds. *Appl. Anim. Behav. Sci.* **2005**, *91*, 75–84. [CrossRef]
17. Coenen, A.M.L.; Lankhaar, J.; Lowe, J.C.; McKeegan, D.E.F. Remote monitoring of electroencephalogram, electrocardiogram, and behavior during controlled atmosphere stunning in broilers: Implications for welfare. *Poult. Sci.* **2009**, *88*, 10–19. [CrossRef] [PubMed]
18. Cornetto, T.; Estevez, I. Behavior of the Domestic Fowl in the Presence of Vertical Panels. *Poult. Sci.* **2001**, *80*, 1455–1462. [CrossRef]
19. Nason, G.P.; Von Sachs, R.; Sachs, R.V. Wavelets in time-series analysis. *Philos. Trans. R. Soc. A Math. Phys. Eng. Sci.* **1999**, *357*, 2511–2526. [CrossRef]
20. Percival, D.B.; Walden, A.T. *Wavelet Methods for Time Series Analysis*; Cambridge University Press: Cambridge, UK, 2000; Volume 4.
21. Nourani, V.; Alami, M.T.; Aminfar, M.H. A combined neural-wavelet model for prediction of Ligvanchai watershed precipitation. *Eng. Appl. Artif. Intell.* **2009**, *22*, 466–472. [CrossRef]
22. Fontana, I.; Tullo, E.; Butterworth, A.; Guarino, M. An innovative approach to predict the growth in intensive poultry farming. *Comput. Electron. Agric.* **2015**, *119*, 178–183. [CrossRef]
23. Bowling, D.L.; Garcia, M.; Dunn, J.C.; Ruprecht, R.; Stewart, A.; Frommolt, K.-H.; Fitch, W.T. Body size and vocalization in primates and carnivores. *Sci. Rep.* **2017**, *7*, 41070. [CrossRef]
24. Bradbury, J.W.; Vehrencamp, S.L. *Principles of Animal Communication*; Sinauer Associates: Sunderland, MA, USA, 1998.
25. Ginovart-Panisello, G.J.; Alsina-Pagès, R.M.; Sanz, I.I.; Monjo, T.P.; Prat, M.C. Acoustic Description of the Soundscape of a Real-Life Intensive Farm and Its Impact on Animal Welfare: A Preliminary Analysis of Farm Sounds and Bird Vocalisations. *Sensors* **2020**, *20*, 4732. [CrossRef]
26. Clapham, W.M.; Fedders, J.M.; Beeman, K.; Neel, J.P. Acoustic monitoring system to quantify ingestive behavior of free-grazing cattle. *Comput. Electron. Agric.* **2011**, *76*, 96–104. [CrossRef]
27. Rau, L.M. Developments on real-time monitoring of grazing cattle feeding behavior using sound. In Proceedings of the 2020 IEEE International Conference on Industrial Technology (ICIT), Buenos Aires, Argentina, 26–28 February 2020.
28. Chelotti, J.O.; Vanrell, S.R.; Milone, D.H.; Utsumi, S.A.; Galli, J.R.; Rufiner, H.L.; Giovanini, L.L. A real-time algorithm for acoustic monitoring of ingestive behavior of grazing cattle. *Comput. Electron. Agric.* **2016**, *127*, 64–75. [CrossRef]
29. Zhang, T.; Wang, J. Design and Implementation of Cow Chewing Behavior Recognition Based on Sound Sensor. In *Data Processing Techniques and Applications for Cyber-Physical Systems (DPTA 2019)*; Springer: Berlin/Heidelberg, Germany, 2020; pp. 817–822.
30. Meluzzi, A.; Sirri, F. Welfare of broiler chickens. *Ital. J. Anim. Sci.* **2009**, *8*, 161–173. [CrossRef]
31. Odén, K.; Keeling, L.; Algers, B. Behaviour of laying hens in two types of aviary systems on 25 commercial farms in Sweden. *Br. Poult. Sci.* **2002**, *43*, 169–181. [CrossRef]

animals

MDPI

Article

Analysis of Cluster and Unrest Behaviors of Laying Hens Housed under Different Thermal Conditions and Light Wave Length

Aline Mirella Fernandes [1], Diogo de Lucca Sartori [2], Flávio José de Oliveira Morais [2], Douglas D'Alessandro Salgado [2] and Danilo Florentino Pereira [1,3,*]

1 Graduate Program in Agribusiness and Development, School of Science and Engineering, São Paulo State University, Tupã 17602-496, Brazil; alinemirellafernandes@gmail.com

2 Department of Biosystems Engineering, School of Science and Engineering, São Paulo State University, Tupã 17602-496, Brazil; diogo.sartori@unesp.br (D.d.L.S.); flavio.morais@unesp.br (F.J.d.O.M.); douglas.salgado@unesp.br (D.D.S.)

3 Department of Management, Development and Technology, School of Science and Engineering, Sao Paulo State University, Tupã 17602-496, Brazil

* Correspondence: danilo.florentino@unesp.br

Simple Summary: The effects of lighting wavelength on the behavior of laying hens are not yet completely known. This study observed three groups of birds housed under different lighting colors (blue, green, and red) for 90 days. Important differences were found regarding the unrest and cluster behaviors of the birds. It was found that, at shorter wavelengths (blue light), birds became more agitated, while, at longer wavelengths (red light), birds became more clustered. When subjected to cold or heat stress, birds expressed unrest and cluster behaviors in different ways, indicating that further studies should be conducted to better clarify the effects of lighting on the behavior and well-being of laying hens.

Abstract: Laying hens are affected by the intensity, wavelength, and duration of light, and the behavioral patterns of these animals are important indicators of stress. The objective of the present study was to evaluate cluster and unrest behaviors of lying hens submitted to three environments with different treatments of monochromatic lighting (blue, green, and red). For 29 weeks, 60 laying hens from the Lohmann variety were divided into three groups and monitored by surveillance cameras installed on each shed ceiling and directed to the floor. Each group was housed in a small-scale shed and maintained under a monochromatic lighting treatment. The recordings were made at two times of the day, 15 min in the morning and 15 min in the afternoon, and the videos were processed, segmented, and analyzed computationally. From the analysis of the images, the cluster and unrest indexes were calculated. The results showed the influence of lighting on these behaviors, displaying that the birds were more agitated in the treatments with shorter wavelengths. Cluster behavior was higher in birds housed under red light. There was an interaction between the lighting treatments and the thermal environment, indicating that more studies should be carried out in this area to better understand these behavioral changes.

Keywords: image analysis; precision poultry farming; animal welfare; movement analysis; LED; comfort index

Citation: Fernandes, A.M.; de Lucca Sartori, D.; de Oliveira Morais, F.J.; Salgado, D.D.; Pereira, D.F. Analysis of Cluster and Unrest Behaviors of Laying Hens Housed under Different Thermal Conditions and Light Wave Length. *Animals* **2021**, *11*, 2017. https://doi.org/10.3390/ani11072017

Academic Editors: Yang Zhao, María Cambra-López, Daniella Jorge De Moura and Weichao Zheng

Received: 30 April 2021
Accepted: 30 June 2021
Published: 6 July 2021

1. Introduction

The use of artificial lighting in the breeding of laying hens is essential to achieve the necessary illuminance, spectrum of light, and suitable photoperiod for the physiological stimulation of the animals [1]. Thus, lighting has a great influence on the productivity of these animals and is a factor of high importance for the welfare of birds confined in conventional egg production systems.

Well-being is defined as the animal's ability to interact and live well in its environment [2]. Light is an essential factor in the microclimate of the poultry house, which means variations in its distribution, intensity, wavelength, and duration affect the welfare and performance of birds [3–7].

The lighting period can contribute to normal and healthy behavior patterns [4,8]. Broilers are more active when in contact with high illuminance (180–200 lux) [9]. For broiler chickens, longer wavelengths (orange and red) make these birds more agitated and aggressive [4].

Behavior is an indicator of stress, and it is affected by the wavelength of light [10–17]. Broilers move more under long-wavelength lighting and tend to stay seated and stationary for longer in environments with blue and green light [5]. Aggressive behavior can be controlled by decreasing the light intensity or using different wavelengths [18]. However, knowledge about the effects of different wavelengths on laying hens is still limited [7].

Birds have four types of single cones, double cones, and rods [19]. Olsson and colleagues [20] report that single-cone photoreceptors are responsible for color vision, each sensitive to a range of specific wavelengths. The maximum sensitivity of these cones is for long wavelengths (L, red) 571 nm, medium wavelengths (M, green) 508 nm, short wavelengths (S, blue) 455 nm, and very short wavelengths (VS, ultraviolet) 415 nm [21]. When a thermal environment changes from thermoneutrality to heat or cold stress, the behavior of birds, whether individual or collective, occurs more quickly in order to mitigate its effects [22–24].

When birds are subjected to heat stress situations, several changes in energy metabolism start to occur, altering thermoregulatory and behavioral responses, and part of the energy that would be used for egg production is redirected to maintain the bird's homeostasis [12,25]. In this situation, one can observe increased water intake, reduced feed consumption, increased respiratory rate, and behaviors such as aggressive pecking and wing exposure as a way to dissipate endogenous heat and maintain homeostasis [26–29].

The behavioral observation of animals can be performed by a human being present at the place where the animals are housed. However, this is a time-consuming, expensive, subjective, and error-prone method. Automated monitoring, through digital cameras, has the ability to generate data that provide an objective measure of behavior, without disturbing animals [30]. In addition to being a low-cost technology, it enables the monitoring of animal behavior on an automated [31–33], non-invasive [34], and ongoing basis.

Digital cameras have been used to monitor the behavior of birds, in which the images analyzed use computer vision techniques [32,35–37]. Computer vision is responsible for extracting relevant information based on images captured by digital cameras, whether through photographic or video images, sensors, and other devices [38]. These technologies have shown great evolution over the past few years [39].

The cluster behavior of laying hens can be classified automatically through image analysis [34]. Pereira and co-workers [34] found that, in conditions of lower temperatures, the laying hens agglomerate more, suggesting that this group behavior can be used to estimate bird thermal comfort. The evaluation of laying hens' agitation behavior was proposed by [40], through an unrest index calculated from image analysis. This index was used to estimate bird thermal comfort, and the authors found that, in high temperature conditions, birds moved less in the poultry house. The combined use of these methods can contribute to a more accurate assessment of the conditions and well-being of commercial birds at their breeding place.

The objective of this work was to evaluate the cluster and unrest behaviors of laying hens in different thermal conditions (cold, comfort, and heat), submitted to three different monochromatic lighting sources (blue, green, and red) in order to verify whether the wavelength of the light source influences these behaviors.

2. Materials and Methods

The study was carried out at the facilities of the Bioterium of the School of Sciences and Engineering, from the São Paulo State University (UNESP), in the city of Tupã, Brazil. The experiments were carried out for 90 days, from 10 June to 8 September 2020, in which the first seven days were dedicated to the adaptation of the birds to the new environment and accommodation conditions.

2.1. Description of Birds and Facilities

For this study, 60 laying hens of the Lohmann variety were monitored at, initially, 29 weeks of age. At the beginning of the experiments, the birds, which were obtained from a commercial farm, were randomly divided into three groups of 20 birds each. Food was administered daily, in the amount of 110 g/bird, once a day, in the morning. Access to water was ad libitum, through nipple drinkers. The light management was similar to that adopted by the original farm, with a photoperiod of 17 h of light.

Three models of sheds were used on a reduced scale, arranged in an east–west orientation, where the birds were housed in a 15 cm high shavings bed. Each poultry house had two 40 × 40 × 40 cm^3 box-type nests installed, a pendular feeder (Φ 30 cm), and four nipple drinkers, as shown in Figure 1a.

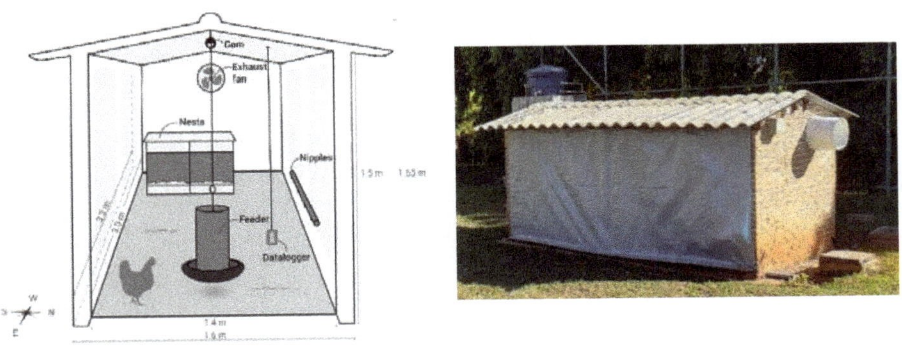

(a) (b)

Figure 1. Experimental houses used in the study: (**a**) layout of the nests, feeders and drinking fountains inside the sheds; (**b**) external view of an sheds.

The sheds were completely sealed with the use of plastic sheeting, to prevent external lighting from influencing the internal lighting system of each environment. For ventilation control and air renewal inside the sheds, exhaust fans (Φ 30 cm, 120 W, and outflow 30 m^3/min) were installed on one of the longitudinal walls (Figure 1b).

In each of the sheds, the group of birds was exposed to the treatment of monochromatic blue, green, and red LED lighting. (Initially, the experimental design provided for a control treatment, where 20 birds were housed under white light. However, in this treatment there was an outbreak of *Lipeurus caponis* lice that affected the behavior of the birds, and the data could not be used in this work). Each light source had a different light spectrum, as shown in Figure 2. The number of lamps in each treatment was calculated from the characteristics of the lamps provided by the manufacturer, to provide the same 100 lux illuminance, similar to the methodology used by Zupan [41].

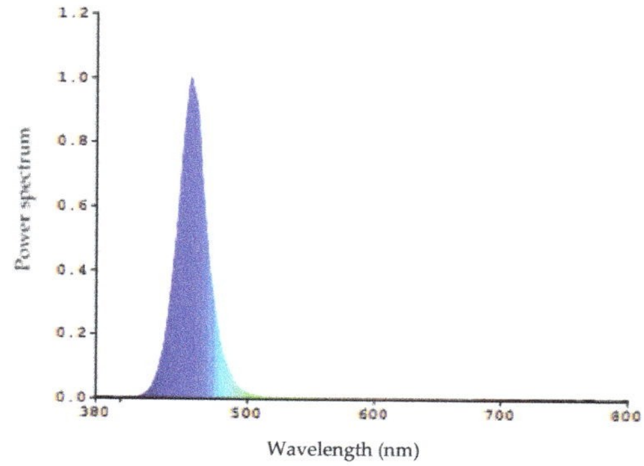

(a)

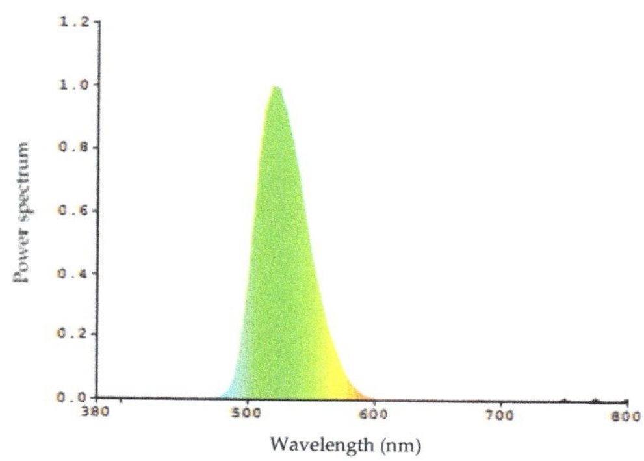

(b)

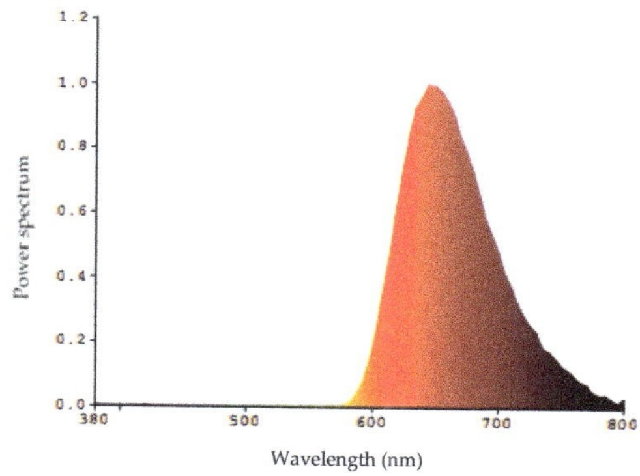

(c)

Figure 2. Luminous spectra of the lamps used in the experiment: (**a**) Model TKL Colors—Blue; (**b**) Model TKL Colors—Green; (**c**) Model TKL Colors—Red. (Data provided by the manufacturer).

2.2. Monitoring of the Thermal Environment

In order to monitor the internal thermal environment of the sheds, a datalogger of the HOBO® brand, model U12-012, was installed. The datalogger was positioned in the center of the sheds, at the same height as that of the birds, to record data on temperature and relative humidity. This equipment was programmed to record the temperature and relative humidity of the air at 5 min intervals for 24 h throughout the entire observational study.

From the temperature and relative humidity data, the Temperature and Humidity Index (*THI*) was calculated for each shed internal environment, using Equation (1), described by [42] and used for birds by [43].

$$THI = 0.8 \times T + \frac{RH \times (T - 14.3)}{100} + 46.3 \quad (1)$$

where: T = temperature dry bulb in °C; RH = relative humidity of the air (%).

2.3. Bird Monitoring System

The birds were monitored by digital surveillance cameras, which were installed in the center of the shed ceiling and directed to the floor, at a 1.5 m height. The cameras recorded for 15 min in the morning and 15 min in the afternoon, according to the methodology used by [44,45]. The captured images were recorded and stored in video format using Digital Video Recorder (DVR) equipment.

The transmission of the images from the cameras to the DVR was made by coaxial cables. The video cameras installed were from the POWER® brand, model AP2688W, with a Charge-Coupled Device (CCD) analog image sensor. The resolution was that of 352 × 240 pixels, a lens with a focal length of 2.8 mm, a viewing angle of 60°, and video standard NTSC (National Television System(s) Committee). The DVR equipment was the model VD 4E120 of the Intelbras® brand, with the Linux operating system installed, supported video format NTSC, had a video recording speed of 30 frames per second (fps) and with capacity for 4 video channels and support for 1 1TB SATA HD.

From the framing obtained by the cameras inside each shed, an area free of objects and equipment was defined, so that the activity of the birds could be monitored. This area was delimited in the first frame and replicated for all consecutive frames of all the video files that composed the samples. The images were processed and analyzed using MATLAB® software. In the image processing, low-pass filters and threshold-based segmentation techniques were applied so that only the birds were highlighted (Figure 3).

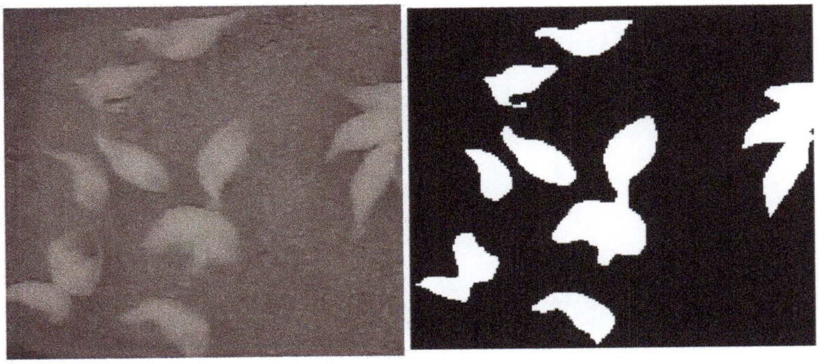

(a) (b)

Figure 3. Example of segmentation of an image: (**a**) Original cropped image; (**b**) segmented image.

From the segmented (binarized) images, it was possible to extract measures used to calculate the cluster and unrest indexes, used in this work to describe the group behavior of the birds in each treatment.

2.4. Measures of Cluster and Unrest Indexes

Two indicators were used to describe the group behavior of birds: the cluster index, described by [37] and the unrest index, described by [40]. The cluster behavior is characterized by the reduction in distances between individuals and the pillaging of these birds.

On the other hand, the unrest behavior is associated with the movement of the flock in the experimental house.

With the segmented image, the positions of the centers of mass of the birds (or groups of birds) were recorded in each video frame, in addition to the area and the perimeter of the shapes that the birds assume. These measures were used to calculate the cluster index described by Equation (2).

$$Cluster\ Index_i = \frac{2 \times \overline{A} \times \sqrt{h^2 + w^2}}{\overline{P} \times \overline{D} \times n_A} - 1 \quad (2)$$

where: $Cluster\ Index_{(i)}$ is the cluster index of the birds observed in the ith frame of the video; $\overline{A}$ and $\overline{P}$ are, respectively, the average area and perimeter (in pixels) of the shapes observed in the frame; $\overline{D}$ is the average distance between the centers of mass of the shapes in the scene; n_A is the number of clusters, and h and w correspond to height and width (in pixels) of the cropped image.

For the calculation of the unrest index (measured in centimeters), initially, the distances from the birds' centers of mass in one frame, at time $i-1$, were calculated to the birds' centers of mass in the next frame, at time i. From the distance measurements between the centers of mass of the birds between the frames, the Hausdorff distance was extracted, which is the mathematical measure that represents the distance between two sets. The Hausdorff distance makes up the unrest index, as described by Equation (3).

$$Unrest\ Index_{(i,i-1)} = k.max\left\{dH\left(F_{(i)}, F_{(i-1)}\right),\ dH\left(F_{(i-1)}, F_{(i)}\right)\right\} \quad (3)$$

where: $Unrest\ Index_{(i,\ i-1)}$ is the unrest index (cm) of the birds between two frames recorded with 1 (one) second difference; i is the position of the frame in the video; $F_{(i)}$ is the current frame; $F_{(i-1)}$ is the previous frame; dH is the Hausdorff distance between group of birds from one frame to the other, and k is the proportionality factor calculated by Equation (4).

$$k = \frac{2H \tan(\alpha/2)}{w} \quad (4)$$

where: k is the proportionality factor; H is the height (cm) of the installed camera in relation to the floor; α is the opening angle of the camera lens, and w is the length (pixels) of the CCD sensor, which corresponds to the length of the largest measurement of the frame captured by the camera.

2.5. Analysis

This is considered an observational cohort study, as it followed three groups of similar individuals (cohorts) under different environmental treatments. Treatments were under blue, green, or red lighting conditions in each experimental house.

In this study, agitation and agglomeration behaviors were compared using the unrest index and cluster index, respectively. Initially, exploratory analyses were performed through graphical interpretations, and later confirmatory analyses through the analysis of variance and the multiple means comparisons test.

3. Results and Discussion

3.1. Thermal Environment

For laying hens, it is considered that temperature and THI values above 28 °C and 78, respectively, are considered situations in which the birds are outside the thermal comfort zone and, therefore, already characterize heat stress [43]. On the other hand, temperatures below 15 °C and a THI below 59 are considered to induce cold stress [46].

Figure 4 shows the variation of the THI for each hour during the entire period of the experiment. When the THI values of the environment are below or above the thermoneutrality limits defined in the literature, the values are highlighted in red.

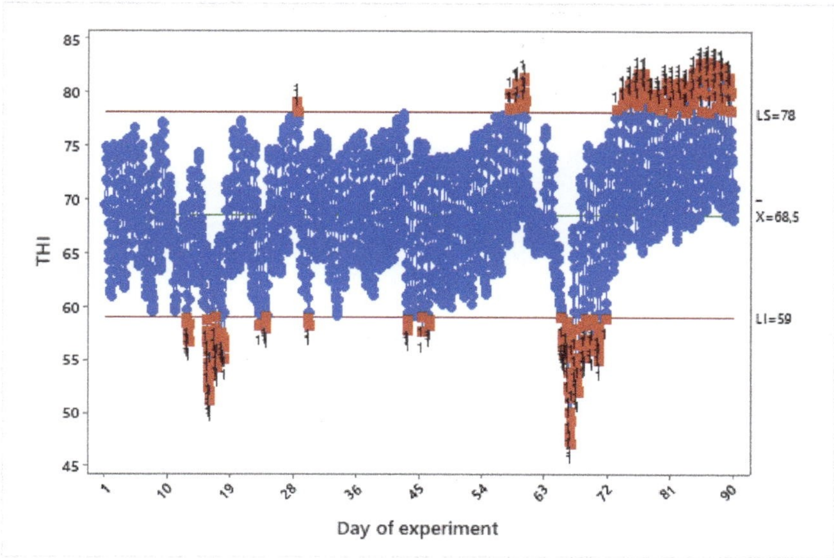

Figure 4. THI averages calculated for each hour, during the period of the observational study.

Figure 4 shows that the birds were exposed to conditions of a thermoneutral environment most of the time. However, they were also exposed to conditions of cold and heat stress at some moments. Considering the recording times of the videos for behavior analysis, 49 thermoneutrality recordings, 20 heat recordings, and 15 cold recordings were obtained.

3.2. Behavior Analysis

Approximately 36 h of images were recorded in a video file in the three lighting treatments, which provided an analysis of about 520,000 frames, allowing the assessment of birds' unrest and cluster behaviors through their respective indexes.

It was observed that the birds' unrest decreases with increasing wavelength (Figure 5a). The group of birds housed under blue lighting treatment were those that showed greater unrest behavior, compared to the other treatments. The confidence intervals for cluster behaviors between the lighting treatments, where the influence of the red wavelength is verified in the increase in the intensity of this collective behavior, are shown in Figure 5b.

For broilers, Sultana and co-workers [5] and Hesham and colleagues [47] found that the birds clustered less under blue lighting (short wavelength) and showed greater unrest when exposed to red lighting (long wavelength). In this study, laying hens exposed to red light were more crowded and less agitated when compared to green and blue lights. The results suggest that the effects of lighting wavelength promote different effects in broilers and laying hens, as also noted by Wichman and colleagues [48], or that age or sexual maturity are determinants for the choice of which light spectrum is the most suitable for each stage of production, as verified by Wei and co-workers [7] in breeding commercial poultry. Red monochromatic LED lighting reduced aggression [49] and reduced bird mortality [50], indicating that this wavelength may be associated with reduced stress. In broilers, studies have shown that, under red light, the birds are more agitated and aggressive [4], while laying hens have an increase in egg production [51] and reduction in stress [52].

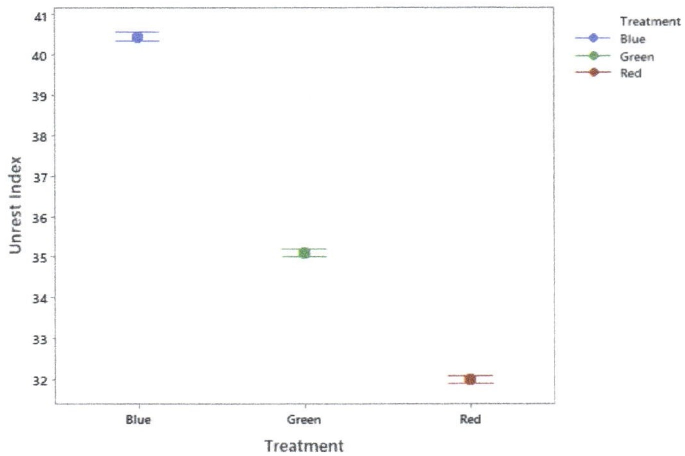

(a)

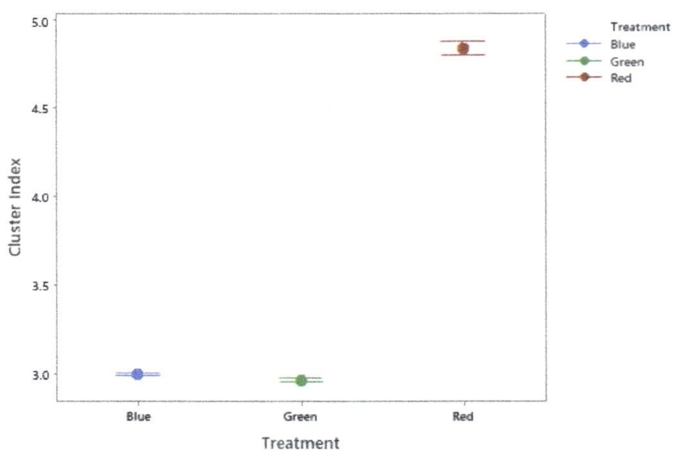

(b)

Figure 5. Confidence intervals of the mean for the (**a**) unrest index and (**b**) cluster index verified for the blue, green, and red light treatments.

Marino [53] describes that personality is defined by three traits (boldness, activity/exploration, and vigilance) and that bird emotions are a combination of cognitive ability and sociability. Birds are highly dependent on vision to express behaviors, especially social behaviors [54,55]. Thus, in environments with monochromatic lighting, it is expected that visual acuity is affected and that social behaviors are altered, influencing the exploration behavior and the welfare of the birds. During the experiment, it was noticeable that the birds under blue lighting were more agitated, as shown by the unrest index (Figure 6a), followed by the green and red treatments. Despite some interaction between the days, there is a tendency to reduce the unrest with the increase in the wavelength.

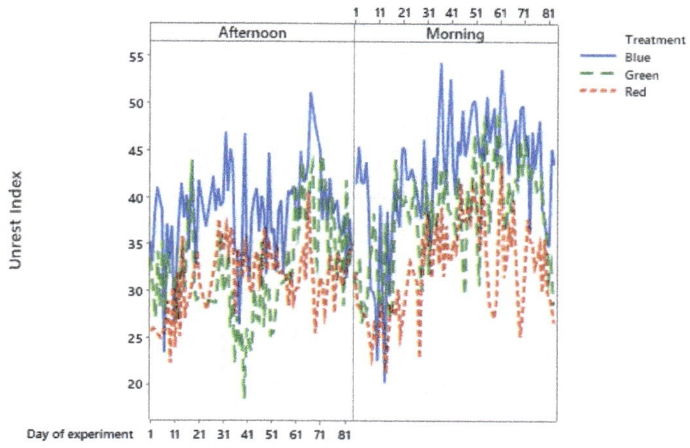

(**a**)

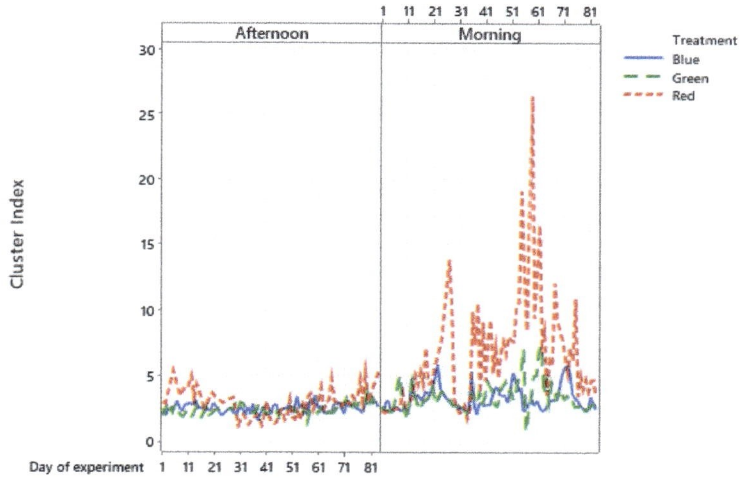

(**b**)

Figure 6. Average daily values for (**a**) unrest index and (**b**) cluster index, referring to samples of 15 min videos recorded and analyzed by period of the day.

In the birds' cluster behavior, a greater interaction of the cluster index was observed between the data from the blue and green light groups. However, the cluster in the group housed under red light was much greater, showing very pronounced peaks in the morning period (Figure 6b). Early in the day, there was a greater supply of food in the feeders and, revisiting the videos, it was found that this cluster occurred around the feeder, demonstrating that birds at this wavelength are more willing to eat.

Lighting is known to affect the behavior of birds [13,15–17]. The birds eat more when exposed to green light when compared to blue light [16]. The birds spend more time around the drinker when under blue and white light, and less time under red and green light [11]. In this work, the dwelling times in the feeder and drinker were not monitored, but the results suggest that there was a greater cluster of birds observed in the red wavelength around the feeder. Birds prefer environments with short-wave lighting (blue and green) to environments with red lighting [14]. Chickens perceive light at a different intensity than humans [56]. Photoreceptors have colored oil droplets that act as a filter for different wavelengths of light [20]. For this reason, each photoreceptor is sensitive to different wavelengths range, with violet light photoreceptors being the most sensitive, followed by blue, green, and red, in that order [19]. In this experiment, this characteristic of the birds' vision may have affected the laying hens' behaviors in response to the light intensity, so that, under the blue light treatment, the birds may have been hyper-stimulated, which would explain the more agitated behavior.

The interaction between the unrest and cluster behaviors of the birds was verified for the thermal conditions observed in the study (Figure 7).

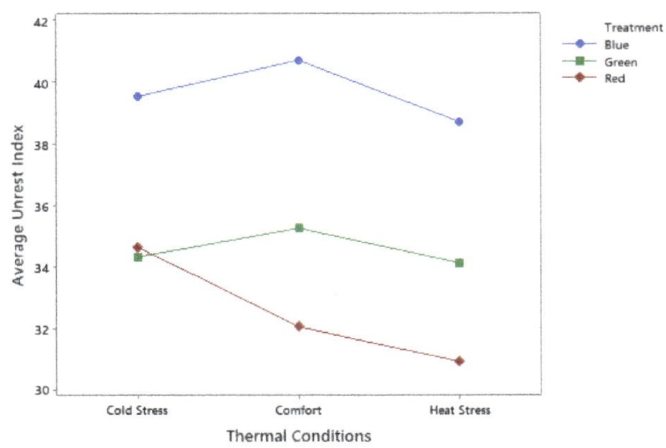

(**a**)

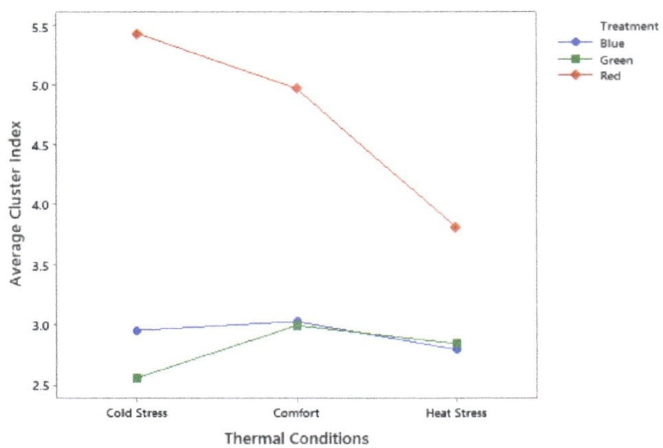

(**b**)

Figure 7. Interaction for (**a**) unrest and (**b**) cluster behaviors in relation to comfort, according to treatment type.

While it is observed that birds reared under red light show a reduction in the unrest index as the temperature increases, birds reared under green and blue light showed a higher unrest index when the temperature was of thermal comfort. Although the results indicate this interaction, it can be seen in Figure 7a that that there is an influence of blue light in the greatest unrest in birds, for all environmental conditions, followed by green and red light, in that order. This figure also shows that, in the blue and green treatments, the unrest is greater in thermoneutrality. Under red light, there was a decrease in unrest with the increase in the wavelength, showing that light wavelength can affect the behavioral response of birds to thermal stress.

Figure 7b reinforces the evidence that the red light influences a greater willingness of birds to feed in all the thermal conditions observed in this study. There is also a tendency to reduce cluster under heat stress conditions in all treatments.

In Tables 1 and 2, the means comparison test was applied for the crossover results of lighting treatment for periods of the day and thermal environments in which the unrest and cluster behaviors were evaluated. The results confirm the evidence presented in the previous figures.

Table 1. Unrest index for each treatment between the levels of the variables: period of the day, production cycle, and thermal comfort.

		Treatment		
		Blue	**Green**	**Red**
Period of day	Morning	40.0 ^{Aa}	34.7 ^{Ab}	32.0 ^{Bc}
	Afternoon	38.3 ^{Ba}	31.8 ^{Bc}	33.4 ^{Ab}
Comfort	Cold Stress	38.1 ^{Ba}	31.2 ^{Bc}	35.2 ^{Ab}
	Comfort	40.7 ^{Aa}	35.3 ^{Ab}	31.5 ^{Bc}
	Heat Stress	38.6 ^{Ba}	33.2 ^{Ab}	31.4 ^{Bc}

Lowercase letters $^{(a,b)}$ indicate differences between the lighting treatments (columns) and uppercase letters $^{(A,B)}$ indicate differences between the lines of the same variable, by the Tukey test at 5% significance.

Table 2. Cluster index for each treatment between the levels of the variables: period of the day, production cycle, and thermal comfort.

		Treatment		
		Blue	**Green**	**Red**
Period of day	Morning	3.16 ^{Ab}	3.15 ^{Ab}	6.18 ^{Aa}
	Afternoon	2.82 ^{Bb}	2.54 ^{Bb}	2.96 ^{Ba}
Comfort	Cold Stress	3.07 ^{b}	2.64 ^{Bc}	4.66 ^{Aa}
	Comfort	2.98 ^{b}	2.92 ^{Ac}	4.79 ^{Aa}
	Heat Stress	2.91 ^{b}	2.98 ^{Ab}	4.26 ^{Ba}

Lowercase letters $^{(a,b)}$ indicate differences between the lighting treatments (columns) and uppercase letters $^{(A,B)}$ indicate differences between the lines of the same variable, by the Tukey test at 5% significance.

Under high temperature conditions, birds clustered less [37] and moved less [12], corroborating the results of this study. Birds prefer to feed in the morning [57,58]. In the afternoon, they remain seated, stationary, for longer periods of time. The results of Table 2 and 3 corroborate these observations, because in the morning, there was a greater cluster of birds around the feeder, for all treatments, while in the afternoon, there was less unrest, except under the red light, where the movement of the birds was higher in the afternoon. The presence of food attracts birds to the feeder and, therefore, increases the cluster of birds around it [57,59,60]. This increase in cluster behavior at the arrival of fresh food is also associated with the common bird behavior of feeding in groups [61].

4. Conclusions

The different monochromatic lighting regimes affected bird behaviors of unrest and cluster. It was found that the unrest was greater under blue light, followed by green and red, which indicates that the increase in the wavelength of the light source may be associated with a lower level of expression of the unrest behavior, or even that longer wavelengths have a calming effect on laying hens. However, studies with more birds are needed to prove this hypothesis.

The interaction was verified between the lighting treatments and the thermal environment, suggesting that further studies should be carried out in this area to better under-stand these behavioral changes.

Author Contributions: The conceptualization was made by A.M.F., D.F.P., D.d.L.S. and F.J.d.O.M.; the methodology was written by A.M.F., D.F.P. and D.d.L.S.; the investigation was made by A.M.F., D.F.P. and F.J.d.O.M.; data curation was carried out by A.M.F. and D.F.P.; the statistical analysis was carried out by D.F.P. and D.D.S.; the original draft was written by A.M.F. and D.F.P.; proofreading and editing of the manuscript was carried out by D.F.P. and D.d.L.S.; visualization and supervision was made under D.F.P., D.d.L.S. and F.J.d.O.M. This study is part of the master's thesis of A.M.F., under the supervision of D.F.P., D.d.L.S. and F.J.d.O.M. All authors have read and agreed to the published version of the manuscript.

Funding: This study was financed in part by the Coordination for the Improvement of Higher Education Personnel (CAPES, Brazil)—Finance Code CAPES/DS Process No. 88881.593696/2020-01.

Institutional Review Board Statement: This study was conducted according to the guidelines of the Brazilian National Council for the Control of Animal Experimentation (CONCEA) and approved by the Sao Paulo State University's Animal Ethics Committee (protocol number 02/2020, CEUA-UNESP), of Comfort Environmental Laboratory at the School of Science and Engineering from the Sao Paulo State University (UNESP, Brazil).

Data Availability Statement: Data will be available upon request to the corresponding author.

Conflicts of Interest: The authors declare no conflict of interest.

References

1. Barros, J.S.G.; Barros, T.A.S.; Morais, F.J.O.; Sartor, K.; Rossi, L.A. Proposal of LED-based linear lighting systems with low power consumption and high light distribution for laying hens. *Comp. Elect. Agric.* **2020**, *169*, 105218. [CrossRef]
2. Broom, D.M. Animal welfare, concepts and measurement. *J. Anim. Sci.* **1991**, *69*, 4167–4175. [CrossRef]
3. Olanrewaju, H.A.; Miller, W.W.; Maslin, W.R.; Collier, S.D.; Purswell, J.L.; Branton, S.L. Effects of light sources and intensity on broilers grown to heavy weights. Part 1, Growth performance, carcass characteristics, and welfare indexes. *Poult. Sci.* **2016**, *95*, 727–735. [CrossRef]
4. Soliman, F.N.K.; El-Sabrout, K. Light wavelengths/colors, Future prospects for broiler behavior and production. *J. Vet. Behav.* **2020**, *36*, 34–39. [CrossRef]
5. Sultana, S.; Hassan, R.; Choe, H.S.; Ryu, K.S. The effect of monochromatic and mixed LED light colour on the behaviour and fear responses of broiler chicken. *Av. Biol. Res.* **2013**, *6*, 207–214. [CrossRef]
6. Janczak, A.M.; Riber, A.B. Review of rearing-related factors affecting the welfare of laying hens. *Poult. Sci.* **2015**, *94*, 1454–1469. [CrossRef] [PubMed]
7. Wei, Y.; Zheng, W.; Li, B.; Tong, Q.; Shi, H. Effects of a two-phase mixed color lighting program using light-emitting diode lights on layer chickens during brooding and rearing periods. *Poult. Sci.* **2020**, *99*, 4695–4703. [CrossRef] [PubMed]
8. Bessei, W. Welfare of broilers, a review. *Worlds Poult. Sci. J.* **2006**, *62*, 455–466. [CrossRef]
9. Blatchford, R.; Klasing, K.C.; Shivaprasad, H.L.; Wakenell, P.S.; Archer, G.S.; Mench, J.A. The effect of light intensity on the behavior, eye and leg health, and immune function of broiler chickens. *Poult. Sci.* **2009**, *88*, 20–28. [CrossRef]
10. Jong, I.C.; Hindle, V.A.; Butteworth, A.; Engel, B.; Ferrari, P.; Gunnink, H.; Perez Moya, T.; Tuyttens, F.A.M.; van Reenen, C.G. Simplifying the Welfare Quality® assessment protocol for broiler chicken welfare. *Animal* **2016**, *10*, 117–127. [CrossRef]
11. Li, G.; Ji, B.; Li, B.; Shi, Z.; Zhao, Y.; Dou, Y.; Brocato, J. Assessment of layer pullet drinking behaviors under selectable light colors using convolutional neural network. *Comp. Elect. Agric.* **2020**, *172*, 105333. [CrossRef]
12. Saeed, M.; Abbas, G.; Alagawany, M.; Kamboh, A.A.; El-Hack, M.E.A.; Khafaga, A.F.; Chao, S. Heat stress management in poultry farms, A comprehensive overview. *J. Therm. Biol.* **2019**, *84*, 414–425. [CrossRef]
13. Er, D.; Wang, Z.; Cao, J.; Chen, Y. Effect of Monochromatic Light on the Egg Quality of Laying Hens. *J. Appl. Poult. Res.* **2007**, *16*, 605–612. [CrossRef]
14. Li, G.; Li, B.; Zhao, Y.; Shi, Z.; Liu, Y.; Zheng, W. Layer pullet preferences for light colors of light-emitting diodes. *Animal* **2019**, *13*, 1245–1251. [CrossRef] [PubMed]
15. Mendes, A.S.; Reffati, R.; Restelatto, R.; Paixão, S.J. Visão e iluminação na avicultura moderna. *Rev. Bras. Agroc.* **2010**, *16*, 5–13. [CrossRef]
16. Silva, G.F.; Tavares, B.O.; Pereira, D.F. Behavior of laying hens with different monochromatic light sources. *Rev. Bras. Eng. Biossistemas* **2012**, *6*, 148–158. [CrossRef]

17. Tsutsui, K.; Ubuka, T.; Bentley, G.E.; Kriegsfeld, L.J. Gonadotropin-inhibitory hormone (GnIH), Discovery, progress and prospect. *Gen. Comp. Endoc.* **2012**, *177*, 305–314. [CrossRef]
18. Olanrewaju, H.A.; Thaxton, J.P.; Dozier, W.A.; Purswell, J.; Roush, W.B.; Branton, S.L. A Review of Lighting Programs for Broiler Production. *Int. J. Poult. Sci.* **2006**, *5*, 301–308. [CrossRef]
19. Seifert, M.; Baden, T.; Osorio, D. The retinal basis of vision in Chicken. *Semin. Cell Dev. Biol.* **2020**, *106*, 106–115. [CrossRef]
20. Olsson, P.; Wilby, D.; Kelber, A. Spatial summation improves bird color vision in low light intensities. *Vis. Res.* **2017**, *130*, 1–8. [CrossRef]
21. Yoshizawa, T. The road to color vision: Structure, Evolution and function of chicken and gecko visual pigments. *Photochem. Photobiol.* **1992**, *56*, 859–867. [CrossRef] [PubMed]
22. Costa, L.S.; Pereira, D.F.; Bueno, L.G.F.; Pandorfi, H. Some Aspects of Chicken Behavior and Welfare. *Braz. J. Poult. Sci.* **2012**, *14*, 159–232. [CrossRef]
23. Oliveira, D.L.; Nascimento, J.W.B.; Camerini, N.L.; Silva, R.C.; Furtado, D.A.; Araujo, T.G.P. Performance and quality of egg laying hens raised in furnished cages and controlled environment. *Rev. Bras. Eng. Agríc. Amb.* **2014**, *18*, 1186–1191. [CrossRef]
24. Vercese, F.; Garcia, E.A.; Sartori, J.R.; Silva, A.P.; Faitarone, A.B.G.; Berto, D.A.; Molino, A.B.; Pelícia, K. Performance and Egg Quality of Japanese Quails Submitted to Cyclic Heat Stress. *Braz. J. Poult. Sci.* **2012**, *14*, 37–41. [CrossRef]
25. Yousaf, A.; Jabbar, A.; Rajput, N.; Memon, A.; Shahnawaz, R.; Mukhtar, N.; Farooq, F.; Abbas, M.; Khalil, R. Effect of Environmental Heat Stress on Performance and Carcass Yield of Broiler Chicks. *World's Vet. J.* **2019**, *9*, 26–30. [CrossRef]
26. Abbas, A.O.; El-dein, A.K.A.; Desoky, A.A.; Galal, M.A.A. The Effects of Photoperiod Programs on Broiler Chicken Performance and Immune Response. *Int. J. Poult. Sci.* **2008**, *7*, 665–671. [CrossRef]
27. Lara, L.J.; Rostagno, M.H. Impact of Heat Stress on Poultry Production. *Animal* **2013**, *3*, 356–369. [CrossRef]
28. Mack, L.A.; Felver-Gant, J.N.; Dennis, R.L.; Cheng, H.W. Genetic variations alter production and behavioral responses following heat stress in 2 strains of laying hens. *Poult. Sci.* **2013**, *92*, 285–294. [CrossRef] [PubMed]
29. Selye, H. Forty years of stress research, principal remaining problems and misconceptions. *Can. Med. Assoc. J.* **1976**, *115*, 53–56. Available online: https//pubmed.ncbi.nlm.nih.gov/1277062/ (accessed on 1 April 2021). [PubMed]
30. Aydin, A.; Cangar, O.; Eren Ozcan, S.; Bahr, C.; Berckmans, D. Application of a fully automatic analysis tool to assess the activity of broiler chickens with different gait scores. *Comp. Elect. Agric.* **2010**, *73*, 194–199. [CrossRef]
31. Barron, D.G.; Brawn, J.D.; Weatherhead, P.J. Meta-analysis of transmitter effects on avian behaviour and ecology. *Met. Ecol. Evol.* **2010**, *1*, 180–187. [CrossRef]
32. Li, N.; Ren, Z.; Li, D.; Zeng, L. Review, Automated techniques for monitoring the behaviour and welfare of broilers and laying hens, Towards the goal of precision livestock farming. *Animal* **2020**, *14*, 617–625. [CrossRef] [PubMed]
33. Siegford, J.M.; Berezowski, J.; Biswas, S.K.; Daigle, C.L.; Gebhardt-Henrich, S.G.; Hernandez, C.E.; Thurner, S.; Toscano, M.J. Assessing Activity and Location of Individual Laying Hens in Large Groups Using Modern Technology. *Animal* **2016**, *6*, 10. [CrossRef] [PubMed]
34. Astill, J.; Dara, R.A.; Fraser, E.D.G.; Roberts, B.; Sharif, S. Smart poultry management, Smart sensors, big data, and the internet of things. *Comp. Elect. Agric.* **2020**, *170*, 105291. [CrossRef]
35. Saltoratto, A.Y.K.; Silva, F.A.; Camargo, A.C.A.C.; Silva, P.C.G.; Souza, L.F.A. Monitoring of aviculture from computer vision techniques. *Colloq. Exactarum* **2013**, *5*, 47–66. [CrossRef]
36. Dawkins, M.S.; Roberts, S.J.; Cain, R.J.; Nickson, T.; Donnelly, C.A. Early warning of footpad dermatitis and hockburn in broiler chicken flocks using optical flow, bodyweight and water consumption. *Vet. Rec.* **2017**, *180*, 499. [CrossRef]
37. Pereira, D.F.; Lopes, F.A.L.; Gabriel Filho, L.R.A.; Salgado, D.D.; Neto, M.M. Cluster index for estimating thermal poultry stress (gallus gallus domesticus). *Comp. Elect. Agric.* **2020**, *177*, 105704. [CrossRef]
38. Shapiro, L.; Stockman, G. *Computer Vision*; Prentice Hall: Hoboken, NJ, USA, 2001.
39. Rajan, A.J.; Jayakrishna, K.; Vignesh, T.; Chandradass, J.; Kannan, T.T.M. Development of computer vision for inspection of bolt using convolutional neural network. *Mat. Today Proc. J.* **2021**, in press. [CrossRef]
40. Del Valle, J.E.; Pereira, D.F.; Mollo Neto, M.; Gabriel Filho, L.R.A.; Salgado, D.D. Unrest index for estimating termal conforto f poultry birds (Gallus gallus domesticus) using computer vision techniques. *Biosyst. Eng.* **2021**, *206*, 123–134. [CrossRef]
41. Zupan, M.; Kruschwitz, A.; Huber-Eicher, B. The influence of light intensity during early exposure to colours on the choice of nest colours by laying hens. *Appl. Anim. Behav. Sci.* **2007**, *105*, 154–164. [CrossRef]
42. Pires, M.F.A.; Ferreira, A.M.; Saturnino, H.M.; Teodoro, R.L. Gestation rate of Holstein females confined in free stall, during the summer and winter. *Arq. Bras. Med. Vet. Zoot.* **2002**, *54*, 57–63. [CrossRef]
43. Biaggioni, M.A.M.; Mattos, J.M.; Jasper, S.P.; Targa, L.A. Thermal performance in layer hen house with natural acclimatization. *Semina Ciências Agrárias* **2008**, *29*, 961–972. Available online: https//repositorio.unesp.br/handle/11449/5178 (accessed on 1 April 2021). [CrossRef]
44. Bizeray, D.; Estevez, I.; Leterrier, C.; Faure, J.M. Effects of increasing environmental complexity on the physical activity of broiler chickens. *Appl. Anim. Behav. Sci.* **2002**, *79*, 27–41. [CrossRef]
45. Pereira, D.F.; Batista, E.S.; Sanches, F.T.; Gabriel Filho, L.R.A.; Bueno, L.G.F. Behavioral differences of laying hens reared in different thermal environments. *Energ. Agric.* **2015**, *30*, 33–40. [CrossRef]
46. Tinôco, I.F.F. A granja de frangos de corte. In *Produção de Frangos de Corte*; Mendes, A.A., Nääs, I.A., Macari, M., Eds.; FACTA: Campinas, Brasil, 2004.

47. Hesham, M.; El Shereen, A.; Enas, S. Impact of different light colors in behavior, welfare parameters and growth performance of Fayoumi broiler chickens strain. *J. Hell Vet. Med. Soc.* **2018**, *69*, 1–9. [CrossRef]
48. Wichman, A.; de Groot, R.; Håstad, O.; Wall, H.; Rubene, D. Influence of Different Light Spectrums on Behaviour and Welfare in Laying Hens. *Animals* **2021**, *11*, 924. [CrossRef] [PubMed]
49. Huber-Eicher, B.; Suter, A.; Spring-Stähli, P. Effects of colored light-emitting diode illumination on behavior and performance of laying hens. *Poult. Sci.* **2013**, *92*, 869–873. [CrossRef] [PubMed]
50. Svobodová, J.; Tůmová, E.; Popelářová, E.; Chodová, D. Effect of light colour on egg production Effect of light colour on egg production. *Czech J. Anim. Sci.* **2015**, *60*, 550–556. [CrossRef]
51. Çapar Akyüz, H.; Onbaşilar, E.E. Light Wavelength on Different Poultry Species. *World's Poult. Sci. J.* **2017**, *74*, 79–88. [CrossRef]
52. Archer, G.S. Animal Well-Being and Behavior How Does Red Light Affect Layer Production, Fear, and Stress? *Poult. Sci.* **2019**, *98*, 3–8. [CrossRef]
53. Marino, L. Thinking chickens: A review of cognition, emotion, and behavior in the domestic Chicken. *Anim. Cogn.* **2017**, *20*, 127–147. [CrossRef] [PubMed]
54. Prescott, N.B.; Wathes, C.M. Spectral sensitivity of the domestic fowl (Gallus g. domesticus). *Br. Poult. Sci.* **1999**, *40*, 332–339. [CrossRef] [PubMed]
55. Collins, S.; Forkman, B.; Kristensen, H.H.; Sandoe, P.; Hocking, P.M. Investigating the importance of vision in poultry: Comparing the behaviour of blind and sighted chickens. *Appl. Anim. Behav. Sci.* **2011**, *133*, 60–69. [CrossRef]
56. Lewis, P.D.; Morris, T.R. Poultry and Coloured Light. *World's Poult. Sci. J.* **2000**, *56*, 203–207. [CrossRef]
57. Barbosa Filho, J.A.D.; Silva, I.J.O.; Silva, M.A.N.; Silva, C.J.M. Behavior evaluation of laying hens using image sequences. *Eng. Agríc.* **2007**, *27*, 93–99. [CrossRef]
58. Farias, M.R.S.; Leite, S.C.B.; Vasconcelos, A.M.; Silva, T.A.G.; Leitão, A.M.F.; Sena, T.L.; Pacheco, D.B.; Abreu, C.G.; Silveira, R.M.F. Thermoregulatory, behavioral and productive responses of laying hens supplemented with different types and dosages of phytases raised in a hot environment, An integrative approach. *J. Therm. Biol.* **2020**, *94*, 102773. [CrossRef] [PubMed]
59. Fernández, A.P.; Norton, T.; Tullo, E.; Hertem, T.V.; Youssef, A.; Exadaktylos, V.; Vandren, E.; Guarino, M.; Berckmans, D. Real-time monitoring of broiler flock's welfare status using camera-based technology. *Biosyst. Eng.* **2018**, *173*, 103–114. [CrossRef]
60. Sirovnik, J.; Würbel, H.; Toscano, M.J. Feeder space affects access to the feeder, aggression, and feed conversion in laying hens in an aviary system. *Appl. Anim. Behav. Sci.* **2018**, *198*, 75–82. [CrossRef]
61. Oliveira, J.L.; Xin, H.; Wu, H. Impact of feeder space on laying hen feeding behavior and production performance in enriched colony housing. *Animal* **2018**, *13*, 374–383. [CrossRef]

animals

MDPI

Article

New Insights into the Hourly Manure Coverage Proportion on the Manure Belt in a Typical Layer House for Accurate Ammonia Emission Modeling

Li Yang [1,†], Chaowu Yang [1,†], Chenming Hu [1], Chunlin Yu [1], Siyang Liu [1], Shiliang Zhu [1], Mohan Qiu [1], Hongqiang Zhu [2], Lingzhi Xie [3] and Longhuan Du [3,*]

[1] Sichuan Animal Science Academy, Chengdu 610066, China; yangli_sasa@163.com (L.Y.); cwyang@foxmail.com (C.Y.); huchenming@126.com (C.H.); yuchunlin1984@sina.com (C.Y.); uniyaliu@163.com (S.L.); zhushiliang1994@163.com (S.Z.); mohan.qiu@163.com (M.Q.)
[2] Sichuan Shengxing Intelligent Technology Company, Chengdu 611436, China; shxgroupzhu@163.com
[3] Department of Mechanics, College of Architecture and Environment, Wangjiang Compus, Sichuan University, Chengdu 610065, China; xielingzhi@scu.edu.cn
* Correspondence: longhuan_du@163.com
† These authors contributed equally to this work.

Simple Summary: Hourly manure coverage proportion and area on the manure belt are key parameters for estimating ammonia emissions in poultry houses in order to provide environmental control suggestions and achieve the goals of precision poultry farming. In this study, experimental measurements were performed, and binary images were applied to provide new insights into the projected hourly manure coverage area on the manure belt at different layer hen ages. It was demonstrated that manure coverage proportion and area measured at different laying hen ages showed similar trends and values with four distinct stages within 48 h. In addition, statistical analyses found no significant correlation between the hourly increment of manure weight and the hourly increment of manure coverage proportion. The results from the present study are expected to serve as a fundamental input parameter for ammonia emission modeling to more accurately simulate the hourly indoor environment and provide effective mitigation strategies.

Abstract: The main advantage of having livestock, for example, the laying hens, in a controlled environment is that the optimum growth conditions can be achieved with accuracy. The indoor air temperature, humidity, gases concentration, etc., would significantly affect the animal performance, thus should be maintained within an acceptable range. In order to achieve the goals of precision poultry farming, various models have been developed by researchers all over the world to estimate the hourly indoor environmental parameters so as to provide decision suggestions. However, a key parameter of hourly manure area in the poultry house was missing in the literature to predict the ammonia emission using the recently developed mechanistic model. Therefore, in order to fill the gap of the understanding of hourly manure coverage proportion and area on the manure belt, experimental measurements were performed in the present study using laying hens from 10 weeks age to 30 weeks age. For each test, six polypropylene (pp) plates were applied to collect the manure dropped by the birds every hour, and photographs of the plates were taken at the same time using a pre-fixed camera. Binary images were then produced based on the color pictures to determine the object coverage proportion. It was demonstrated that for laying hens of stocking density around 14 birds/m^2, the manure coverage proportion at the 24th hour after the most recent manure removal was about 60%, while the value was approximately 82% at the 48th hour. Meanwhile, for laying hens at different ages, the hourly increment of manure coverage proportion showed a similar pattern with four distinct stages within 48 h. The statistical analyses demonstrated no significant correlation between the hourly increment of manure weight and the hourly increment of manure coverage proportion. Finally, prediction models for estimating the hourly manure coverage proportion on the manure belt in typical laying hen houses were provided.

Citation: Yang, L.; Yang, C.; Hu, C.; Yu, C.; Liu, S.; Zhu, S.; Qiu, M.; Zhu, H.; Xie, L.; Du, L. New Insights into the Hourly Manure Coverage Proportion on the Manure Belt in a Typical Layer House for Accurate Ammonia Emission Modeling. *Animals* **2021**, *11*, 2433. https://doi.org/10.3390/ani11082433

Academic Editors: Riccardo Bozzi and Maria Caria

Received: 13 July 2021
Accepted: 17 August 2021
Published: 18 August 2021

Keywords: manure area; manure coverage proportion; environment control; ammonia emission; layer house

1. Introduction

In recent decades, the intensification of poultry production in China has contributed to ensuring increasing demand for domestic livestock. Small farms with traditional systems have been replaced by controlled environment housings. Poultry production in an enclosed environment with high stocking density becomes an important source of ammonia (NH_3) emission, which has a number of negative effects not only on the indoor air quality but also on the ecosystem [1–3]. For typical poultry production, the ammonia originates from the decomposition of nitrogen content in manure and the production and emission of the NH_3 are a result of complex biological, physical, and chemical processes [4,5]. Moreover, various factors, including ventilation rate, temperature, humidity, stocking density, management, etc., would affect the indoor ammonia concentration and emissions [6–8].

The ammonia concentration in controlled environment housings should be kept within an acceptable range [9,10] since a high concentration of NH_3 had been demonstrated to be associated with health risks for both birds and exposed workers [11,12]. Therefore, it is crucial to understand and model the ammonia emissions in poultry houses so as to provide information to develop appropriate mitigation and management strategies. Much work had been done to predict ammonia release from manure [13–17], and several types of models were developed in the literature, including statistical models [18,19], balance models [20,21], and process-based models [22,23]. More recently, Tong et al. [24] developed a mechanistic model, which was based on the fundamental understanding of physical and biochemical processes of ammonia emissions from manure, to estimate the NH_3 emissions rate (ER, mg $m^{-2}h^{-1}$) from laying hen manure. Information including manure pH, manure moisture content (MC), air velocity, air temperature, etc., were required for the model, and readers could refer to the original paper for more detailed information. The total ammonia emissions (M_{NH3}, mg h^{-1}) could then be calculated by $M_{NH3} = ER \times A_s$, where A_s was the manure surface area, m^2.

Knowing the parameter of A_s, the above mechanistic model could be effectively incorporated into many recently developed thermodynamic models [25–27], which were used to predict the indoor hourly environmental parameters, including ammonia emissions, and provide decision suggestions in order to achieve the goal of precision poultry farming. Nevertheless, a review of published literature demonstrates that very limited information is available relating to A_s for laying hen production. Considering the difficulty for accurately measuring the hourly A_s, researchers in the literature applied the manure projected area, A_p, on the manure belt to approximate the A_s. According to a recent study performed by Tong et al. [28], the manure coverage proportion (MCP) on the manure belt per day, or more specifically, the coverage proportion of projected manure area on the manure belt per day, was estimated by the equation $MCP = \min\left\{\frac{1}{3} + \frac{d-1}{3}, 1\right\}$, where d was the number of days after manure removal, $\min\{a, b\}$ equaled the smaller value between a and b. Based on the above equation, the daily manure coverage proportion was estimated to be $MCP_{day1} = 33.3\%$, $MCP_{day2} = 66.7\%$, $MCP_{day3} = 100\%$. Unfortunately, to the best of the authors' knowledge, there is no hourly data of manure coverage proportion or A_s available, which could be directly applied for the thermodynamic models for predicting hourly ammonia emissions.

Therefore, this study aimed to fill the research gap by providing new insights into the hourly manure coverage proportion and manure area on the manure belt in a typical layer house. The weekly manure pH and MC was also measured, which were important information for estimating NH_3 emissions. Although it is noted that the hourly manure coverage proportion on the manure belt might, to some extent, be affected by diet, species, stocking density, etc., the results from the present study are expected to serve as a fundamental input parameter for thermodynamic models to more accurately simulate the hourly indoor environment and provide effective management strategies.

2. Materials and Methods

2.1. *The Layer House and the Birds*

The experimental measurements were conducted in an experimental-oriented manure belt layer house in Chengdu, Sichuan province. The dimensions of the house were length, 40 m, width, 9.2 m, height, 2.5 m. Tunnel ventilation is applied with evaporative cooling systems in the house, and more details about the building could be found in previous studies [29,30]. In the house, there were 4 rows of animal-occupied zone. Each row had 3 tiers of cages raising approximately 3500 birds of the parent stock of the local species characterized by partridge-like plumage and dark-shanks. A total of 8 birds were kept in each cage with a size of width 660 mm and length 860 mm, resulting in a stocking density of approximately 14 birds/m^2.

2.2. *Manure Collection*

Pure white polypropylene (pp) plates, which had the same width of the manure belt, 680 mm (slightly larger than the width of the cage) and a length of 860 mm (equal to the length of the cage), were hung above the manure belt in order to collect the manure dropped from the birds as schematically drawn in Figure 1. The polypropylene plates were weighted every hour so as to calculate the updated weight of the manure, and plan-view photographs of the plates with manure were also taken at the same time to determine the updated manure coverage proportion, which would be detailed in Section 2.3. In this study, the measurement campaign was conducted once a week, starting from the laying hen 10 weeks age to 30 weeks age. Meanwhile, for each measurement campaign, 6 polypropylene plates were applied and placed randomly in the poultry house providing enough data (manure produced by 48 laying hens) to calculate the hourly average values. In each week, the test began at 5 am in the morning (lights on) and lasted for 48 h (2 days). Detailed information on the measurement campaign is summarized in Table 1 below.

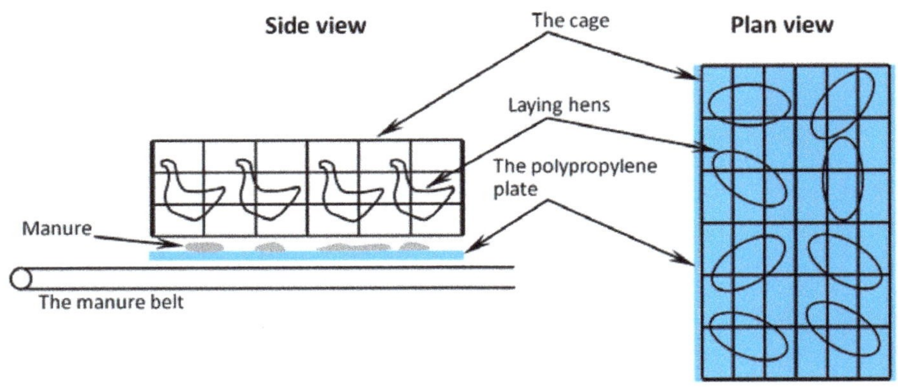

Figure 1. Schematic drawing of the size and placement of the polypropylene plate. The plate length is 860 mm, and the plate width is 680 mm.

Table 1. Detailed information on the experimental measurements.

	Information	Notes
Experiment period	10 weeks age~30 weeks age	Measurements performed once a week
Measurement interval	1 h	Starts at 5 am and lasts for 48 h (2 days)
Number of pp plates	6	To calculate the hourly average values
Parameters concerned	Manure weight, coverage proportion, area	The resolution of the scale is 0.1 g, and the resolution of the photographs is 4032 × 3024 pixels
Staff involved	6 people	Rotating schedule

2.3. Determination of the Manure Coverage Proportion (MCP) and Area

To investigate the hourly manure coverage proportion (MCP) and area on the manure belt, the six polypropylene plates were moved to a pre-marked area one by one every hour to have the photographs taken by a pre-fixed camera (Figure 2). Special attention was paid when transferring the plates so as to reduce the movement of manure on the plates, which was inevitable given the fact that the manure was not 'fixed' on the plates. The camera lens was set perpendicular to the surface of the plate, ensuring that all the pictures were taken at the same position, height, orientation, and resolution in order to minimize the experimental error. In the present study, the resolution of the photographs was determined to be 4032 × 3024 pixels, which was demonstrated to be enough for the following study as pictures with more pixels did not show any significant difference in terms of the results of image processing.

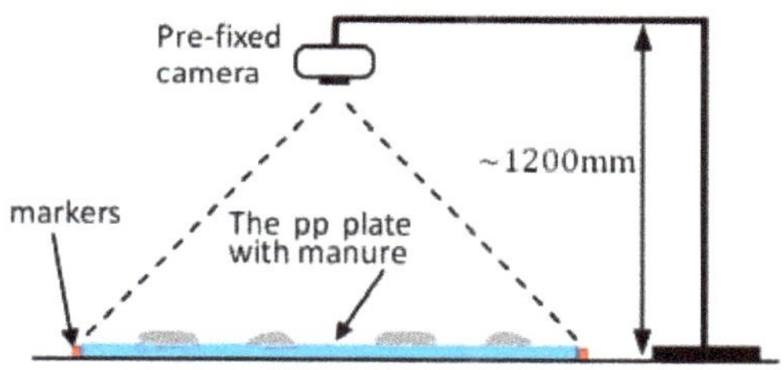

Figure 2. Schematic drawing of the pre-fixed camera.

To be more specific, the manure area investigated in this study was the projected area of the manure on the manure belt. The starting point of how to calculate the projected area from a picture is to estimate the manure coverage proportion in a binary image. As long as the coverage proportion could be determined, the manure area and hourly area increment could be easily calculated since the area of the background polypropylene plate is known. Therefore, the color photographs were firstly turned into gray-scale images in Matlab, and a threshold value, T = 200, was used to check the gray value of each pixel in order to produce binary images, namely, a gray value smaller than 200 would be set to 0 (black) while gray value larger than 200 would be set to 255 (white). Finally, the objects coverage proportion (γ) in the binary image could be easily determined by $\gamma = \frac{\text{number of pixels with gray value}=0}{\text{number of total pixels}} \times 100\%$ and the area could be calculated at the same time. A flowchart is provided in Figure 3 to show the detailed image processes.

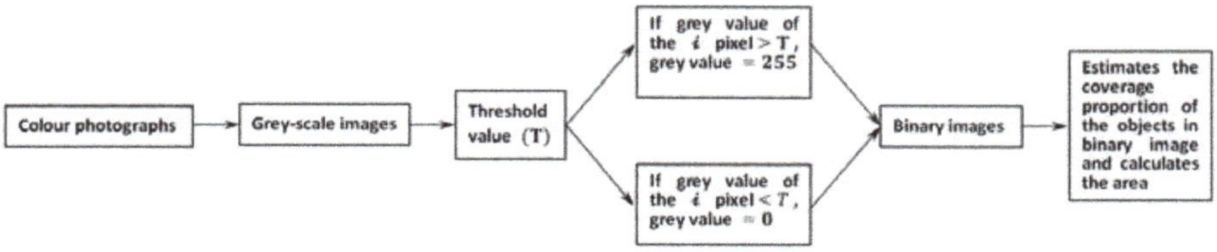

Figure 3. The flowchart for image processing.

Examples of comparisons between original color photographs and binary images are illustrated in Figure 4. As it can be clearly seen from the four pictures in Figure 4, the binary images are capable of replicating almost all details of the manure in color photographs taken at different stages of the experiment, showing the correct position and size. The corresponding manure coverage proportion for Figure 4a–d is calculated to be 11.76%, 25.82%, 36.01%, and 68.63%, respectively, and the corresponding manure area is 0.069 m^2, 0.151 m^2, 0.211 m^2, and 0.401 m^2, respectively. Furthermore, the limited white urate on the manure would, to some extent, affect the accuracy of coverage proportion calculated by binary images, and the maximum discrepancy was investigated to be up to approximately ±3.3% of the estimated coverage proportion value γ.

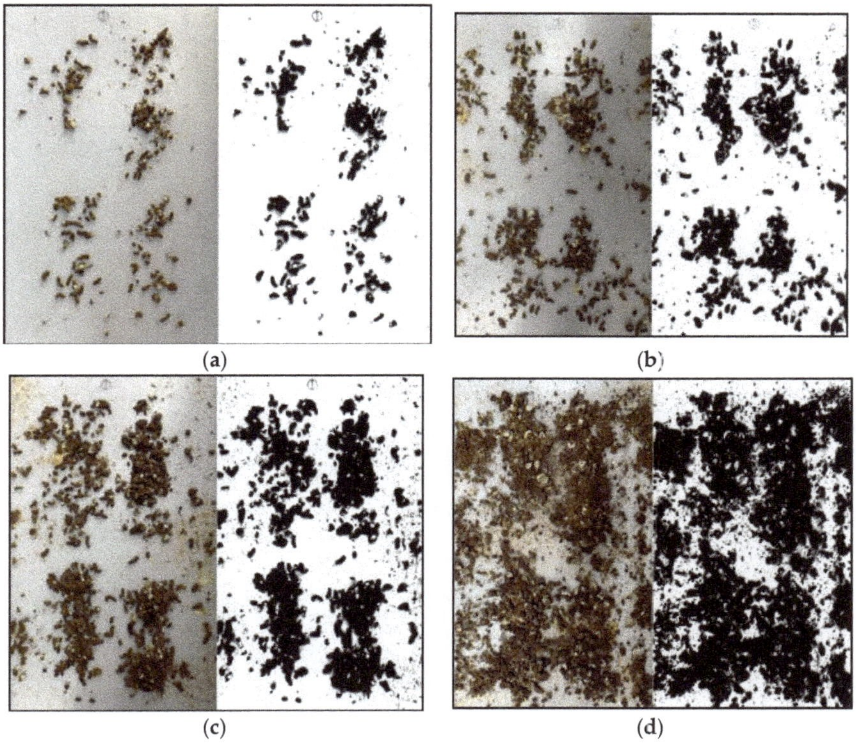

Figure 4. Examples of original photographs (left) and binary images (right). The photographs were taken at the (**a**) 3rd hour, (**b**) 8th hour, (**c**) 11th hour, and (**d**) 30th hour after the start of the experiment. The corresponding objects coverage proportion for (**a**–**c**) and (**d**) is 11.76%, 25.82%, 36.01%, and 68.63%, respectively.

2.4. Manure pH, Moisture Content (MC), and Lighting

For each week during the experiment period, the manure was sampled randomly from multiple locations in the house within 3 h after manure had been dropped by the birds.

The samples were then properly stored in sealed bags and transported in a timely manner to a quality-certified laboratory for determining the pH and MC. Moreover, from the laying hen 10 weeks age to 30 weeks age, the lighting program was modified regularly to achieve optimal reproductive performance through appropriate illumination and photostimulation at the appropriate age and body weight. A detailed lighting schedule for the local species is provided in Table 2.

Table 2. Lighting schedule during experiment period and targeted weight of the hens.

Week	Daylength (Hours)	Targeted Weight (g)	Week	Daylength (Hours)	Targeted Weight (g)
10	13	1140	21	16	2130
11	13	1230	22	16	2220
12	13	1320	23	16	2300
13	14	1410	24	16	2380
14	14	1500	25	17	2460
15	14	1590	26	17	2540
16	14	1680	27	17	2630
17	15	1770	28	17	2700
18	15	1860	29	17	2770
19	15	1950	30	17	2840
20	15	2040			

3. Results and Discussion

3.1. Manure Weight

The hourly increment of manure weight was measured and calculated during each test from the laying hen 10 weeks age to 30 weeks age. The results from four typical ages are presented here in the format of the mean value (M) and standard deviation (ST). As it can be clearly seen in Figure 5, the manure produced by the birds every hour in the daytime (lights on) is apparently more than that in the nighttime (lights off). For 12 weeks age, the average hourly increment of manure weight recorded in the daytime is approximately 7.6 g per hour per hen, while the value is about 3.9 g per hour per hen in the nighttime (see Figure 5a). Furthermore, due to the increase in the amount of feed in the following weeks, the birds produce more manure every hour than that in 12 weeks age. The corresponding average hourly increment of manure weight in the daytime for 18, 24, and 30 weeks age is about 9.2, 11.1, and 11.8 g per hour per hen, respectively. Meanwhile, the corresponding average hourly increment of manure weight recorded in the nighttime for 18, 24, and 30 weeks age is about 5.4, 5.8, and 6.2 g per hour per hen, respectively (see Figure 5).

Although the day length increases gradually from the laying hen 10 weeks age to 30 weeks age (see Table 2) as the laying hens enter the laying period from the rearing period, the recorded feed to manure ratio is kept at around 2.04 in each week as it can be seen in Table 3. Detailed information of average hourly increment of manure weight measured in the daytime and nighttime in each week is also provided in Table 3. In addition, the recorded weekly moisture content (MC) ranges from $72.7\% \pm 4.0\%$ to $82.3\% \pm 2.1\%$, and there is no apparent trend or pattern detected. However, the measured manure PH value demonstrated a downward trend from the beginning of the experiment to the end. The maximum value of $PH = 7.9 \pm 0.3$ is recorded in the 11 weeks age, while the minimum value of $PH = 6.8 \pm 0.1$ is measured in the 28 weeks age. It is hypothesized that the changes in the content of feed and the climate might be responsible for the PH decrease, and further study is required to provide solid conclusions.

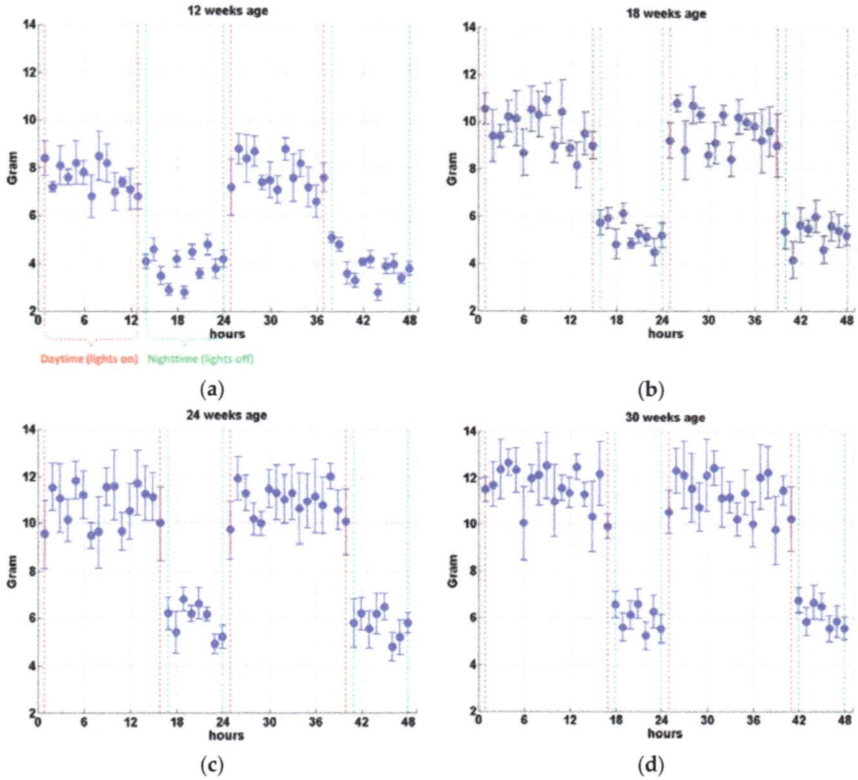

Figure 5. Manure weight measured (per hour per hen) at four typical laying hen ages of (**a**) 12 weeks, (**b**) 18 weeks, (**c**) 24 weeks, and (**d**) 30 weeks. Average value is provided with the standard deviation (error bar). Each test starts at 5 am (lights on) and lasts for 48 h with a measurement interval of one hour. Data between red-dash lines are measured in the daytime (lights on), and data between green-dash lines are recorded in the nighttime (lights off).

Table 3. Manure data measured and recorded each week.

Age (Week)	Feed (g)	Average Hourly Increment of Manure Weight Measured In The Daytime (Gram per Hour per Hen)	Average Hourly Increment of Manure Weight Measured in the Nighttime (Gram per Hour per Hen)	Feed to Manure Ratio	PH (M±ST)	MC, % (M±ST)
10	60	7.0	3.7	2.19	7.5 ± 0.2	73.3 ± 3.9
11	63	7.2	3.9	2.15	7.9 ± 0.3	71.5 ± 2.8
12	67	7.6	3.9	2.12	7.6 ± 0.3	78.7 ± 1.6
13	70	8.0	4.5	2.24	7.3 ± 0.2	79.3 ± 1.1
14	74	8.2	4.6	2.17	7.1 ± 0.1	76.1 ± 1.9
15	78	8.4	4.8	2.11	7.5 ± 0.1	74.5 ± 3.5
16	83	8.5	5.1	2.05	7.6 ± 0.3	81.6 ± 2.9
17	88	8.9	5.0	2.03	7.3 ± 0.1	75.4 ± 1.8
18	93	9.2	5.4	2.01	7.7 ± 0.3	74.5 ± 3.4
19	98	9.6	5.3	1.95	6.9 ± 0.3	78.9 ± 2.3
20	103	10.1	5.6	1.96	7.3 ± 0.2	81.2 ± 1.7
21	108	10.6	5.8	1.97	7.1 ± 0.2	77.4 ± 3.9
22	110	10.9	5.5	2.01	7.0 ± 0.2	72.7 ± 4.0
23	112	11.0	5.9	1.93	7.3 ± 0.1	77.3 ± 3.3
24	114	11.1	5.8	1.94	7.1 ± 0.3	79.1 ± 1.1
25	116	11.4	6.0	1.95	7.4 ± 0.3	72.9 ± 1.5
26	118	11.6	6.1	2.01	7.2 ± 0.2	82.3 ± 2.1
27	120	11.8	6.0	2.02	7.1 ± 0.2	74.9 ± 2.9
28	120	11.7	6.2	2.01	6.8 ± 0.1	75.8 ± 1.5
29	120	11.9	6.1	2.04	6.9 ± 0.1	76.8 ± 2.4
30	120	11.8	6.2	2.03	7.0 ± 0.2	79.2 ± 2.6

Note: manure PH and MC are provided in the format of mean value (M) ± standard deviation (ST).

3.2. Manure Coverage Proportion (MCP) and Area

Figure 6 illustrates the binary images of the manure on one of the pp plates taken at different times during the experiment for 24 weeks age. The pictures indicate the corresponding manure coverage proportion at the 1st, 4th, 8th, 12th, 30th, and 44th hour is approximately 1.65%, 11.42%, 23.77%, 38.93%, 67.12%, and 81.26%, respectively. Meanwhile, the corresponding projected manure area is calculated to be 0.01 m^2, 0.067 m^2, 0.139 m^2, 0.227 m^2, 0.393 m^2, and 0.475 m^2, respectively. Finally, the total manure area in the poultry house (A_T) could then be estimated using the following equation:

$$A_T = CP_i \times A_{plate} \times N_{cage} \quad (1)$$

where CP_i is the manure coverage proportion at the ith hour after the most recent manure removal, A_{plate} is the pp plate area, which is roughly equal to the cage area, and N_{cage} is the total number of cages in the house.

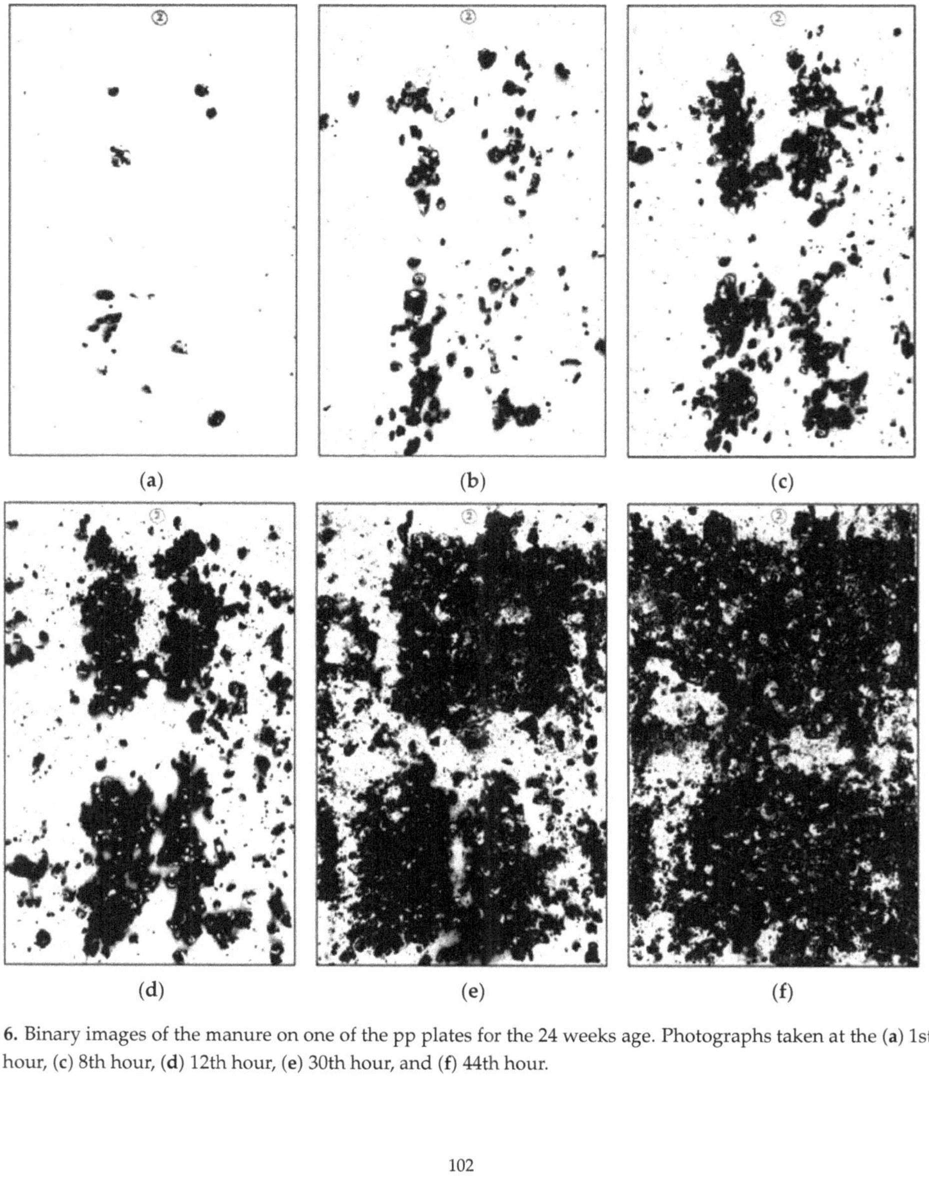

(a) (b) (c) (d) (e) (f)

Figure 6. Binary images of the manure on one of the pp plates for the 24 weeks age. Photographs taken at the (**a**) 1st hour, (**b**) 4th hour, (**c**) 8th hour, (**d**) 12th hour, (**e**) 30th hour, and (**f**) 44th hour.

By observing the photographs of the manure, a clear message can be read that with the increase in coverage proportion, the phenomenon of manure overlap becomes apparent. It is extremely difficult to exactly measure the surface area of the manure due to its irregular shape, but the projected manure area still represents a suitable method to approximate the surface area since the release of ammonia from manure is significantly affected by the airflow characteristics (including temperature, velocity, turbulence, etc.) above the release surface according to previous studies [14]. Therefore, manure covered underneath has a limited contribution to the total NH_3 emissions and the projected area would not be considerably different from the true surface area since the height of the overlap is not large according to the field observation.

Figure 7 further shows the hourly increment of manure coverage proportion (MCP) on the plates measured at four typical laying hen ages. The result is presented in the format of the mean value calculated from six plates with standard deviation (error bar). As it can be seen in Figure 7a, the hourly increment of MCP for the first daytime (from the 1st hour to the 13th hour) is approximately 3.34% per hour. Lights were turned off from the 14th hour for 12 weeks age, and an apparent decrease in hourly increment is recorded in the first nighttime (from the 14th hour to the 24th hour) with a mean value of about 1.35% per hour, which agrees with the decrease in hourly increment of manure weight measured in the nighttime as it can be seen in Figure 5a and Table 3. The total MCP after the first day (the 24th hour) is calculated to be 58.33%, as indicated by the solid black line (right *Y*-axis) in Figure 7a. For the second daytime, the hourly increment of MCP increases at the beginning from the 25th hour to about the 32nd hour due to the feeding activity. However, because of the aggravation of manure overlap, the hourly increment of MCP demonstrates a decreasing trend from the 33rd hour to the 37th hour. The overall mean value of hourly increment of MCP for the second daytime (from the 25th hour to the 37th hour) is measured to be only about 1.43% per hour, which is significantly lower than that in the first daytime although the manure weight dropped by the birds during the second daytime is roughly equal to that during the first daytime as demonstrated in Figure 5a. Because of limited manure dropped by the birds during nighttime and severe manure overlap resulting from the existing large coverage proportion, the hourly increment of MCP measured for the second nighttime (from the 38th hour to the 48th hour) is very limited with an average value of merely 0.32% per hour. Finally, the total manure coverage proportion climbs to about 80.35% at the end of the experiment (the 48th hour).

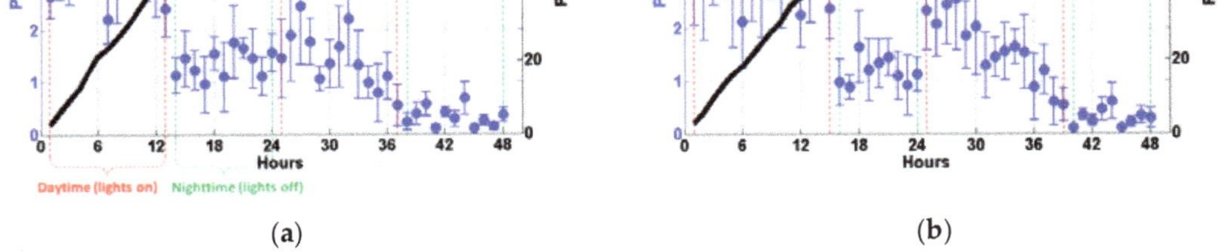

(a) (b)

Figure 7. *Cont.*

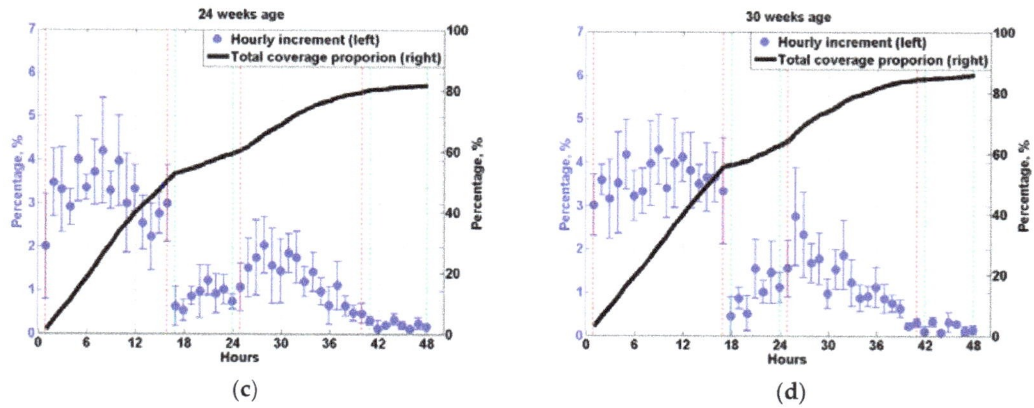

Figure 7. Hourly increment of manure coverage proportion (left *Y*-axis) and total coverage proportion (right *Y*-axis) on the pp plates measured in (**a**) 12 weeks age, (**b**) 18 weeks age, (**c**) 24 weeks age, and (**d**) 30 weeks age. Each test starts at 5 am (lights on) and lasts for 48 h with a measurement interval of one hour. Data between red-dash lines are measured at daytime (lights on), and data between green-dash lines are measured at nighttime (lights off).

As illustrated in Figure 7b, the data measured for laying hen of 18 weeks age shows a similar trend with that in 12 weeks age. The hourly increment of MCP is relatively large at the beginning of the experiment, with a mean value of 3.04% per hour for the first daytime (from the 1st hour to the 15th hour). The average hourly increment decreases to about 1.18% per hour for the first nighttime (from the 16th hour to the 24th hour) in accordance with the decrease in the hourly increment of manure weight (see Figure 5b). For the second daytime from the 25th hour to the 39th hour, the hourly increment of MCP rebounds to approximately 1.61% per hour, which is only about half of that for the first daytime due to the manure overlap. The total manure coverage proportion ends up at about 82.94%, with very limited hourly increments observed from the 40th hour to the 48th hour (the second nighttime).

By examining the results measured in other weeks, for example, the MCP data for 24 and 30 weeks age as shown in Figure 7c,d, it is found that all the recorded coverage proportion curves (the solid black line) demonstrate a similar trend with four distinct stages: firstly, an almost linear relationship is detected between the MCP and the time (hours) with a gradient ranges from about 3.0% to 3.5%; secondly, for the first nighttime the curve slope reduces to about 0.9% $\sim$ 1.4%; thirdly, due to the manure overlap, the total coverage proportion curve during the second daytime only shows a moderate gradient ranges from 1.35% to 1.8%, which is considerably lower than the gradient at the beginning of the test. Finally, when the experiment enters into the second nighttime, the curve gradient reduces to only about 0.3%. The MCP data measured from all the 21 weeks are then averaged, and mean values are plotted in Figure 8. Results from the present study indicate the manure produced by the birds in one day would cover approximately 60% of the area of the manure belt, and more than 80% of the belt area would be covered within 48 h.

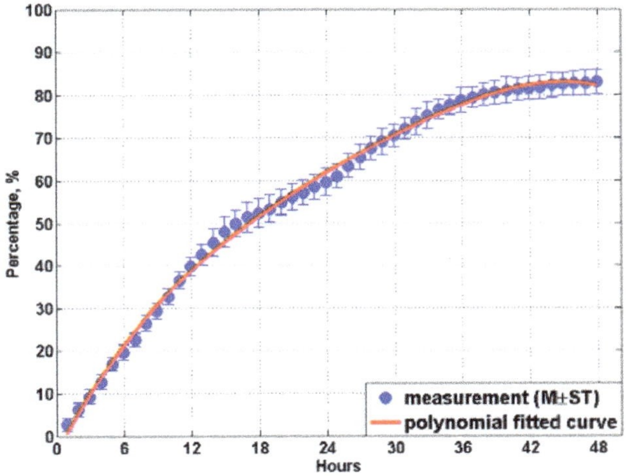

Figure 8. The measured manure coverage proportion curve with time and the polynomial fit curve within 48 h. Mean values (M) and standard deviations (ST) are calculated based on all 21 weeks' data.

Furthermore, no significant correlation ($r = 0.11, p > 0.05$) can be found between the hourly increment of manure weight and the hourly increment of manure coverage proportion by examining the data recorded from 10 to 30 weeks age using Pearson's correlation coefficient in SPSS. The insignificant correlation indicates that the hourly increment of MCP or manure area on the manure belt would not necessarily be affected by the variation of the amount of manure dropped by the birds.

A polynomial fitted curve is created to represent the total manure coverage proportion within 48 h after the most recent manure removal, as can be seen in Figure 8. The fitted curve shows suitable agreement ($R^2 = 0.997$) with the experimental measurements and falls within the standard error range at each hour. The equation of the fitted curve for predicting MCP reads

$$MCP_{48} = P_1 \times h^4 + P_2 \times h^3 + P_3 \times h^2 + P_4 \times h + P_5 \quad (2)$$

where h is the time (hours) after the most recent manure removal and the values of coefficients of $P_1 \sim P_5$ are provided in Table 4.

Table 4. Coefficients for polynomial fitted curves.

Coefficient	Value	Coefficient	Value
P_1	-3.359×10^{-5}	P_6	3.234×10^{-4}
P_2	3.621×10^{-3}	P_7	-1.861×10^{-2}
P_3	-0.1648	P_8	0.2786
P_4	5.081	P_9	1.992
P_5	-4.105	P_{10}	1.147

In addition, for some poultry farms where the manure belt is cleared every 24 h, the equation for predicting MCP within 24 h is also provided and reads

$$MCP_{24} = P_6 \times h^4 + P_7 \times h^3 + P_8 \times h^2 + P_9 \times h + P_{10} \quad (3)$$

where the values of coefficients of $P_6 \sim P_{10}$ is provided in Table 4, and the polynomial fitted curve is shown in Figure 9, which demonstrates suitable agreement ($R^2 = 0.999$) with the data measured from the field tests.

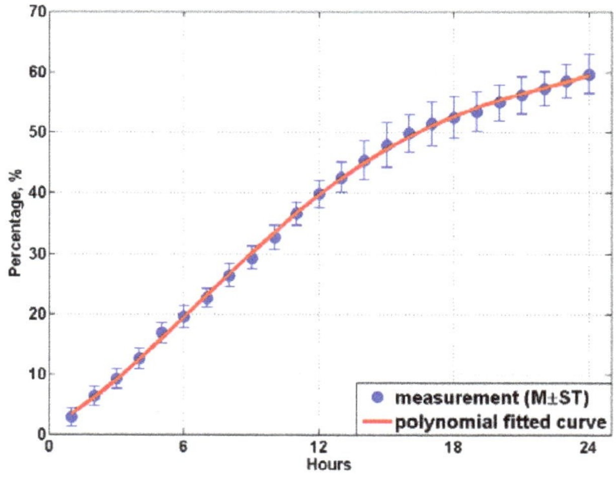

Figure 9. The measured manure coverage proportion with time and polynomial fit curve within 24 h. Mean values (M) and standard deviations (ST) are calculated based on all 21 weeks' data.

4. Conclusions

In order to fill the gap of the understanding of the relationship between manure coverage proportion on the manure belt and time during poultry farming, experimental measurements were performed in a manure belt tunnel-ventilated layer house with a stocking density of about 14 birds/m^2 using laying hens from 10 to 30 weeks age. In each week, the test began at 5 am in the morning and lasted for 48 h with a measurement interval of one hour. Six polypropylene (pp) plates were placed randomly above the manure belt to collect the manure dropped by the hens in order to provide average results. The manure weight was investigated every hour, and photographs of the pp plates were taken at the same time using a pre-fixed camera. Binary images were then produced based on the color pictures, and the objects coverage proportion was estimated, and the manure area was calculated at the same time. In addition, important manure parameters, including pH and moisture content (MC), were also measured every week to provide basic data for future studies.

Based on the experimental results from the present study, some conclusions can be drawn as follows:

1. The feed to manure ratio is kept at $\sim$ 2.04 from the laying hen 10 weeks age to 30 weeks age;
2. The hourly increment of manure coverage proportion measured in different laying hen ages demonstrates similar trends and values with four distinct stages within 48 h;
3. For stocking density around 14 birds/m^2, the manure coverage proportion on the manure belt at the 24th hour after the most recent manure removal is about 60%, while the value is approximately 82% at the 48th hour;
4. The statistical analyses demonstrate no significant correlation between the hourly increment of manure weight and the hourly increment of manure coverage proportion on the manure belt.

Finally, this study provides new knowledge and prediction models for estimating the hourly manure coverage proportion and area in the poultry house, which could be directly applied in thermodynamic models developed in the literature to predict the indoor hourly ammonia emissions achieving the goal of precision poultry farming.

Author Contributions: Conceptualization, L.D.; methodology, L.D. and L.Y.; software, L.D.; investigation, C.Y. (Chaowu Yang), C.H., M.Q., L.Y., C.Y. (Chunlin Yu), S.L., S.Z. and L.D.; resources, H.Z., L.X. and C.Y. (Chaowu Yang); writing—original draft preparation, L.Y., C.Y. (Chaowu Yang) and L.D.; writing—review and editing, L.Y., C.Y. (Chaowu Yang) and L.D. All authors have read and agreed to the published version of the manuscript.

Funding: This work was funded by Sichuan Province Basic Scientific Research Project (SASA202002); Sichuan province key research and development plan (2020YFN0065, 2019YFN0009, 2020YFSY0048, 2021YFN0029,2021YFYZ0031); China Agriculture Research System of MOF and MARA(CARS-41).

Institutional Review Board Statement: Not applicable.

Data Availability Statement: Not applicable.

Acknowledgments: The authors would like to thank the staff in the poultry farm for their kind help with experimental measurements.

Conflicts of Interest: The authors declared that there is no conflict of interest.

References

1. Paulot, F.; Jacob, D.J.; Pinder, R.W.; Bash, J.O.; Travis, K.; Henze, D.K. Ammonia emissions in the United States, European Union, and China derived by high-resolution inversion of ammonium wet deposition data: Interpretation with a new agricultural emissions inventory (MASAGE_NH3). *J. Geophys. Res. Atmos.* **2014**, *119*, 4343–4364. [CrossRef]
2. Xu, R.T.; Pan, S.F.; Chen, J.; Chen, G.S.; Yang, J.; Dangal, S.R.S.; Shepard, J.P.; Tian, H.Q. Half-Century Ammonia Emissions from Agricultural Systems in Southern Asia: Magnitude, Spatiotemporal Patterns, and Implications for Human Health. *GeoHealth* **2018**, *2*, 40–53. [CrossRef]
3. Li, D.; Tong, Q.; Shi, Z.; Zheng, W.; Wang, Y.; Li, B.; Yan, G. Effects of Cold Stress and Ammonia Concentration on Productive Performance and Egg Quality Traits of Laying Hens. *Animals* **2020**, *10*, 2252. [CrossRef]
4. Rotz, C.A. Management to reduce nitrogen losses in animal production. *J. Anim. Sci.* **2004**, *82*, E119–E137.
5. Swelum, A.A.; El-Saadony, M.T.; El-Hack, M.E.A.; Ghanima, M.M.A.; Shukry, M.; Alhotan, R.A.; Hussein, E.O.; Suliman, G.M.; Ba-Awadh, H.; Ammari, A.A.; et al. Ammonia emissions in poultry houses and microbial nitrification as a promising reduction strategy. *Sci. Total Environ.* **2021**, *781*, 146978. [CrossRef]
6. Méda, B.; Hassouna, M.; LeComte, M.; Germain, K.; Dourmad, J.-Y.; Robin, P. Influence of season and outdoor run characteristics on excretion behaviour of organic broilers and gaseous emissions. *Biosyst. Eng.* **2015**, *139*, 35–47. [CrossRef]
7. Li, Y.; Ni, J.-Q. Dynamic and 3-D spatial variations in manure characteristics in two commercial manure-belt laying hen houses. *J. Hazard. Mater.* **2021**, *403*, 123581. [CrossRef]
8. Meda, B.; Hassouna, M.; Aubert, C.; Robin, P.; Dourmad, J. Influence of rearing conditions and manure management practices on ammonia and greenhouse gas emissions from poultry houses. *World's Poult. Sci. J.* **2011**, *67*, 441–456. [CrossRef]
9. Zhao, Y.; Shepherd, T.A.; Li, H.; Xin, H. Environmental assessment of three egg production systems–Part I: Monitoring system and indoor air quality. *Poult. Sci.* **2015**, *94*, 518–533. [CrossRef]
10. Han, H.; Zhou, Y.; Liu, Q.; Wang, G.; Feng, J.; Zhang, M. Effects of Ammonia on Gut Microbiota and Growth Performance of Broiler Chickens. *Animals* **2021**, *11*, 1716. [CrossRef]
11. Ritz, C.W.; Fairchild, B.D.; Lacy, M.P. Implications of Ammonia Production and Emissions from Commercial Poultry Facilities: A Review. *J. Appl. Poult. Res.* **2004**, *13*, 684–692. [CrossRef]
12. Zhou, Y.; Zhang, M.; Liu, Q.; Feng, J. The alterations of tracheal microbiota and inflammation caused by different levels of ammonia exposure in broiler chickens. *Poult. Sci.* **2021**, *100*, 685–696. [CrossRef] [PubMed]
13. Bjerg, B.; Norton, T.; Banhazi, T.; Zhang, G.; Bartzanas, T.; Liberati, P.; Cascone, G.; Lee, I.-B.; Marucci, A. Modelling of ammonia emissions from naturally ventilated livestock buildings. Part 1: Ammonia release modelling. *Biosyst. Eng.* **2013**, *116*, 232–245. [CrossRef]
14. Rong, L.; Nielsen, P.V.; Zhang, G. Experimental and numerical study on effects of airflow and aqueous ammonium solution temperature on ammonia mass transfer coefficient. *J. Air Waste Manag. Assoc.* **2010**, *60*, 419–428. [CrossRef] [PubMed]
15. Rosa, E.; Mosquera, J.; Arriaga, H.; Montalvo, G.; Merino, P. Ammonia emission modelling and reduced sampling strategies in cage-based laying hen facilities. *Biosyst. Eng.* **2021**, *204*, 304–311. [CrossRef]
16. Baker, J.; Battye, W.H.; Robarge, W.; Arya, S.P.; Aneja, V.P. Modeling and measurements of ammonia from poultry operations: Their emissions, transport, and deposition in the Chesapeake Bay. *Sci. Total Environ.* **2020**, *706*, 135290. [CrossRef]
17. Bjerg, B.; Cascone, G.; Lee, I.B.; Bartzanas, T.; Norton, T.; Hong, S.W.; Seo, I.H.; Banhazi, T.; Liberati, P.; Marucci, A.; et al. Modelling of ammonia emissions from naturally ventilated livestock buildings. Part 3: CFD modelling. *Biosyst. Eng.* **2013**, *116*, 259–275. [CrossRef]
18. Wheeler, E.F.; Casey, K.; Gates, R.S.; Xin, H.; Zajaczkowski, J.L.; Topper, P.A.; Liang, Y.; Pescatore, A.J. Ammonia Emissions from Twelve U.S. Broiler Chicken Houses. *Trans. ASABE* **2006**, *49*, 1495–1512. [CrossRef]

19. Liang, W. Ammonia Emission Model for High-Rise Layer Houses. Ph.D. Thesis, Purdue University, West Lafayette, IN, USA, 2015.
20. Chai, L.; Kröbel, R.; Janzen, H.H.; Beauchemin, K.A.; McGinn, S.M.; Bittman, S.; Atia, A.; Edeogu, I.; MacDonald, D.; Dong, R. A regional mass balance model based on total ammoniacal nitrogen for estimating ammonia emissions from beef cattle in Alberta Canada. *Atmos. Environ.* **2014**, *92*, 292–302. [CrossRef]
21. Keener, H.M.; Zhao, L. A modified mass balance method for predicting NH3 emissions from manure N for livestock and storage facilities. *Biosyst. Eng.* **2008**, *99*, 81–87. [CrossRef]
22. Li, C.; Salas, W.; Zhang, R.; Krauter, C.; Rotz, A.; Mitloehner, F. Manure-DNDC: A biogeochemical process model for quantifying greenhouse gas and ammonia emissions from livestock manure systems. *Nutr. Cycl. Agroecosyst.* **2012**, *93*, 163–200. [CrossRef]
23. Pinder, R.W.; Pekney, N.J.; Davidson, C.I.; Adams, P.J. A process-based model of ammonia emissions from dairy cows: Improved temporal and spatial resolution. *Atmos. Environ.* **2004**, *38*, 1357–1365. [CrossRef]
24. Tong, X.; Zhao, L.; Heber, A.J.; Ni, J.-Q. Mechanistic modelling of ammonia emission from laying hen manure at laboratory scale. *Biosyst. Eng.* **2020**, *192*, 24–41. [CrossRef]
25. Costantino, A.; Fabrizio, E.; Ghiggini, A.; Bariani, M. Climate control in broiler houses: A thermal model for the calculation of the energy use and indoor environmental conditions. *Energy Build.* **2018**, *169*, 110–126. [CrossRef]
26. Wang, Y.; Li, B.; Liang, C.; Zheng, W. Dynamic simulation of thermal load and energy efficiency in poultry buildings in the cold zone of China. *Comput. Electron. Agric.* **2020**, *168*, 105127. [CrossRef]
27. Hamilton, J.; Negnevitsky, M.; Wang, X. Thermal analysis of a single-storey livestock barn. *Adv. Mech. Eng.* **2016**, *8*, 1–9. [CrossRef]
28. Tong, X.; Zhao, L.; Heber, A.J.; Ni, J.-Q. Development of a farm-scale, quasi-mechanistic model to estimate ammonia emissions from commercial manure-belt layer houses. *Biosyst. Eng.* **2020**, *196*, 67–87. [CrossRef]
29. Du, L.; Yang, C.; Dominy, R.; Yang, L.; Hu, C.; Du, H.; Li, Q.; Yu, C.; Xie, L.; Jiang, X. Computational Fluid Dynamics aided investigation and optimization of a tunnel-ventilated poultry house in China. *Comput. Electron. Agric.* **2019**, *159*, 1–15. [CrossRef]
30. Du, L.; Yang, L.; Yang, C.; Dominy, R.; Hu, C.; Du, H.; Li, Q.; Yu, C.; Xie, L.; Jiang, X. Investigation of bio-aerosol dispersion in a tunnel-ventilated poultry house. *Comput. Electron. Agric.* **2019**, *167*, 105043. [CrossRef]

animals

MDPI

Article

Implementation of Inertia Sensor and Machine Learning Technologies for Analyzing the Behavior of Individual Laying Hens

Sayed M. Derakhshani [1,2,*], Matthias Overduin [1], Thea G. C. M. van Niekerk [3] and Peter W. G. Groot Koerkamp [1]

1 Farm Technology Group, Wageningen University, 6700 AA Wageningen, The Netherlands; matthiasoverduin@hotmail.com (M.O.); peter.grootkoerkamp@wur.nl (P.W.G.G.K.)
2 Biometris, Wageningen University, 6700 AA Wageningen, The Netherlands
3 Wageningen Livestock Research, Wageningen University, 6700 AA Wageningen, The Netherlands; thea.vanniekerk@wur.nl
* Correspondence: sayed.derakhshani@wur.nl

Simple Summary: Poultry-welfare regulations have caused a shift from cage housing towards more welfare-friendly systems with more possibilities for the birds to meet their natural behavioural needs. The welfare-friendly systems with litter allow and encourage the hens to perform natural behavior including activities that lead to increases in the amount of airborne dust particles emission from such poultry houses. For successful management of these systems, the behavior of the hens needs to be considered, which is more challenging and time-consuming for the farmer. The main objective of this study was to show a proof of principle to identify, classify and analyze the behaviors of laying hens in three levels of activity by using an inertia sensor and machine learning techniques. The model was able to predict the laying hen behaviors with an accuracy of 90%. The results of such monitoring could be used by farmers in the management of poultry houses.

Abstract: Welfare-oriented regulations cause farmers worldwide to shift towards more welfare-friendly, e.g., loose housing systems such as aviaries with litter. In contrast to the traditional cage housing systems, good technical results can only be obtained if the behavior of hens is considered. With increasing flock sizes, the automation of behavioural assessment can be beneficial. This research aims to show a proof of principle of tools for analyzing laying-hen behaviors by using wearable inertia sensor technology and a machine learning model (ML). For this aim, the behaviors of hens were classified into three classes: static, semi-dynamic, and highly dynamic behavior. The activities of hens were continuously recorded on video and synchronized with the sensor signals. Two hens were equipped with sensors, one marked green and one blue, for five days to collect the data. The training data set indicated that the ML model can accurately classify the highly dynamic behaviors with a one-second time window; a four-second time window is accurate for static and semi-dynamic behaviors. The Bagged Trees model, with an overall accuracy of 89% was the best ML model with the F1-scores of 89%, 91%, and 87% for static, semi-dynamic, and highly dynamic behaviors. The Bagged Trees model also performed well in classifying the behaviors of the hen in the validation data set with an overall F1-score of 0.92 (uniform either % or decimals). This research illustrates that the combination of wearable inertia sensors and machine learning is a viable technique for analyzing the laying-hen behaviors and supporting farmers in the management of hens in loose housing systems.

Keywords: laying hen; daily behavior; machine learning; inertia sensor

Citation: Derakhshani, S.M.; Overduin, M.; van Niekerk, T.G.C.M.; Groot Koerkamp, P.W.G. Implementation of Inertia Sensor and Machine Learning Technologies for Analyzing the Behavior of Individual Laying Hens. *Animals* **2022**, *12*, 536. https://doi.org/10.3390/ani12050536

Academic Editors: Yang Zhao, María Cambra-López, Daniella Jorge De Moura and Weichao Zheng

Received: 11 January 2022
Accepted: 18 February 2022
Published: 22 February 2022

1. Introduction

Welfare-oriented legislation, such as the European Directive [1], imposing specific regulations for the keeping of laying hens, has caused a shift from cage housing towards more welfare-friendly systems with more possibilities for the birds to meet their behavioural needs. These so-called loose housing systems typically house larger groups of hens and provide them with a litter area and more space per bird [2]. For successful management of these systems, the behavior of the hens needs to be taken into account, which makes it more challenging for the farmer and also more time-consuming [3].

The shift from cage housing to loose housing systems for laying hens substantially increased the fine-dust emissions from the poultry sector in the Netherlands, with the litter being the major additional source. Fine dust (PM10) from livestock houses consists primarily of faeces, feed and animal matter, such as hairs and feathers [4,5]. Fine dust is regarded as a pollutant that causes harmful effects for both the environment and the health and welfare of humans and animals [6,7]. The emitted PM10 (Particular Matter (PM10) with an aerodynamic equivalent diameter equal to and less than 10 μm [7]) from the poultry sector exceeds the air quality thresholds set by the European Union [8]. The type of activity of laying hens, and in particular the activities in the litter area and their intensity, have a direct and pronounced effect on PM10 emissions. Calvet et al. [9] indicated that there is a strong relationship between the concentration of fine dust in the air and the animal activity index (Ai), demonstrated in a study with three- and four-weeks old broilers. In this case, the activity index was defined as the proportion of broilers that were not laying down. The low Ai during the night and the middle of the day were caused by two dark periods during the day, in which the activity of the broilers was significantly lower.

Various techniques are being developed to reduce dust emission and they affect dust concentration in the animal house and/or in the exhaust air [10]. However, a combination of such techniques with managing bird behavior might enhance the performance of those techniques. Smart managing of light intensities and feeding times can direct birds from or towards the litter [9] and thus regulate and restrict peak dust emission. This would not only benefit the total dust emission but could also benefit the health of the human workers in the house [11].

The daily behavior of production animals needs to be monitored to apply optimal management in poultry houses. Automated monitoring of the behavior of laying hens can have advantages for managing the flock and safeguarding animal welfare [12]. The activity level and type of activity of the birds can provide useful information about individual and flock health (e.g., detect sick birds or piling) [13] and welfare status (e.g., typical positive or negative behaviors). Daily behaviors are not equally distributed over the day, therefore a good impression of behavior can only be obtained if assessed throughout the entire day. As it is too time-consuming for farmers to manually check the chicken behaviors, an automatic device can be very useful. Not all behaviors are equally important for management decisions. Therefore, chicken behaviors can be classified into different classes according to their activity levels. Kozak et al. [14] identified three main classes of laying-hen behavior based on the intensity of their individual activities. They were classified as low-, moderate- and high-intensity physical activities.

In general, animal-monitoring techniques can be divided into the following two categories: body-worn sensor technologies and remote measurement technologies. To monitor chicken behavior, both technologies have advantages and disadvantages. One of the advantages of body-worn sensor technologies is the possibility of the individual identification of chickens, as they all have their own sensor. By using remote measurement technologies, such as computer vision, the distinction between various individual animals can be more difficult. However, in the case of body-worn sensors, one sensor is needed for each chicken, which might cause problems when upscaling the system to a commercial-flock size [15].

Machine learning is a technique that is used in animal behavior analysis studies to efficiently analyze large datasets. This technique consists of a system with multiple algorithms that enable the subtraction of hidden features and relationships from datasets. The complexity of the different algorithms varies and involves several stages of sophisticated decision making, which invites the use of machine learning algorithms into optimizing automating processes [16]. These types of monitoring are even able to operate in real-time, which potentially alleviates the task of monitoring [16].

Machine learning distinguishes two main types of learning, i.e., supervised and unsupervised learning. Supervised learning consists of algorithms that attempt to classify data based on labelled input data, while unsupervised learning models a set of inputs where labelled input is not available. One supervised technique that is often applied in research is the Support Vector Machine (SVM), which is preferred because of its ability in generalization [17]. Hepworth et al. [16] showed that applying SVM techniques to recognize birds/poultry data were able to correctly predict whether a chicken was sick or not (accuracy rate of 99.5%). They also compared additional algorithms such as Bayesian classifier, Random Forest, and an artificial neural network and all these methods had accuracies above 95%.

A review of the literature reveals that there is great potential for improving the performance of these algorithms. Examples of these improvements include classifying more behavioural categories and improving the outcomes of machine learning models by validating the parameters. While accelerometers are widely used, there is no consistent approach to process the data that is being generated by these devices [16]. This restricts the ability to compare results across various research. Additionally, behavioural studies are, for practical reasons, often performed in smaller-sized systems, and results obtained by those studies do not necessarily reflect behavior in commercially sized systems [16]. The main objective of this study was to show a proof of principle of tools for analyzing laying-hen behaviors using wearable inertia sensor technology and machine learning techniques. The outcome of this research will give more insight into the feasibility of using such techniques as part of automation and management systems in the commercial poultry sector. This can aid in managing indoor air quality and dust emission.

2. Materials and Methods

2.1. Experimental Setup

An experiment was designed to collect data of the daily behavior of two individual chickens. The experiment was carried out in a section of a commercial aviary laying-hen house, which was fenced off from the rest of the house by wire mesh. The experimental area was about 5 × 4 × 4 (length-width-height) meters in size. The same facilities were available as in the rest of the house, such as a feed trough, nipple drinkers for unlimited water supply, nest boxes, perches, a pecking block, and a large litter area. The light program was similar to the rest of the house. Lights were turned on around 4:30 a.m. and turned off with a dimming phase between 7:30 p.m. and 7:45 p.m. During the experiment, 15 white laying hens were present in the experimental area. The chickens were 34 weeks of age. Two chickens were equipped with a lightweight inertial measurement unit (IMU, also called inertia sensor) of 16 g. The measurement units were mounted onto the chickens via small backpacks, consisting of a small elastic strap that was looped around the wings, together with a small bag of fabric in which the sensor was placed. Ethical approval for the experiment was granted by the Animal Welfare Body of Wageningen Research for mounting backpacks on two chickens for a maximum period of 5 days.

2.2. Data Acquisition and Analysis

The actual behavior of the chickens was collected simultaneously via video recordings and a wireless inertial sensor. The chicken activities were recorded during the experiment using a GoPro Hero 7 (Black) video camera. The MTw2 Awinda wireless motion tracker (IMU) of Xsens was used as the inertial measurement unit in this study. This sensor consisted of an accelerometer, a gyroscope, and a magnetometer which can measure the acceleration, angular velocity, and magnetic field, respectively. The technical properties for the IMU are provided in Table 1.

Table 1. Technical properties of the MTw2 Awinda Wireless 3DOF Motion Tracker (www.xsens.com, accessed on 8 April 2020).

Parameter	Angular Velocity	Acceleration	Magnetic Field
Dimensions	3 axes	3 axes	3 axes
Full scale	2000 deg/s	160 m/s^2	1.9 Gauss
Non-linearity	0.1% of FS	0.5% of FS	0.1% of FS
Bias stability	10 deg/hr	0.1 mg	-
Noise	0.01 deg/s/$\sqrt{}$Hz	0.01 µg/$\sqrt{}$Hz	0.2 mGauss/$\sqrt{}$Hz
Alignment error	0.1 deg	0.1 deg	0.1 deg
Bandwidth	180 Hz	180 Hz	10–60 Hz (var.)

Two chickens were equipped with a backpack, in which the sensor could be fit. To be able to distinguish one from the other, the color of their backpacks differed, whereby one backpack had a green color mark and the other backpack a blue color mark. The first two days of the experiment were used to get the chickens being used to wearing backpacks. After that, no effects of the backpacks on behavior were expected [18]. During the following three days of data collection, the 'green chicken' wore the sensor in the first two days and the 'blue chicken' wore the sensor during the third day. When a chicken was not wearing a sensor, a foam dummy sensor was placed in the backpack to avoid that the chicken would behave differently in case the sensor was placed in the backpack.

As the sensor was not able to measure the time of the day, a timer was started at the same moment as the sensor was turned on. By displaying this timer in the video recordings, the time of the video and the sensor could be synchronized. The synchronization of the sensor and the video was required because the video was used for the annotation of the behaviors.

Table 2 shows the classification of the behavioural activities of laying hens based on their intensity [16]: low-, moderate- and high-intensity physical activities. Class 1 represents static laying-hen behaviors, class 2 represents semi-dynamic behaviors and class 3 represents highly dynamic laying-hen behaviors.

Table 2. Classification of physical activity of laying hens based on their intensity (adopted and modified from Table 2 in ref [14]).

Class 1	Class 2	Class 3
Low-intensity	Moderate-intensity	High-intensity
• Sleep like resting • Neck shortening resting • Sleeping • Quiet sitting/standing • Small postural head/shoulder/neck movements • Perching • Egg laying • Side-laying phase of dust bathing	• Preening • Foraging & pecking • Drinking & eating • Small wing adjustments • Scratching & stretching • Head shaking • Feather fluffing • Searching behavior • Scratching behavior of dust bathing	• Walking • Running • Jumping • Wing flapping • Controlled aerial ascent/descent • Full-body shaking • Shaking phase of dust bathing

The classes are highly imbalanced in terms of the amount of collected data per class, as class 3 has substantially fewer data points than classes 1 and 2. By assigning higher penalties to minority classes, the ML models can equalize the weights of the classes in case of the existence of an imbalance between available data points. The penalty of misclassifying a sample of class i is calculated by:

$$penalty_i = nTotal/nClass_i\,, \tag{1}$$

where $nTotal$ is the total number of samples in the dataset and $nClass_i$ is the number of samples in the dataset that is annotated as class i [17].

The original dataset [19] consisted of the standardized dataset for the green chicken. Dataset A is a precise version of the original dataset that was redone by a trained person for class 3. Dataset B was used to analyse the classification performance of models obtained from the behavior of the blue chicken. An overview of the different datasets and their class distributions is provided in Table 3.

Table 3. Dataset overview with the available number of data points per class and additional information. Two chickens were observed, one wearing a green backpack and one wearing a blue backpack.

	Original Dataset	Dataset A	Dataset B
Number of datapoints class 1	3023	3017	747
Number of datapoints class 2	3606	3588	2638
Number of datapoints class 3	37	61	47
Total number of data points	6666	6666	3432
Chicken (color)	Green	Green	Blue
Day of recording	Wednesday	Wednesday	Friday
Total length of the recording	2 h 20 min	2 h 20 min	29 min

As seen in Figure 1, classes were highly imbalanced based on the amount of annotated data per class. Hence, penalty matrices were applied to the datasets to compensate for the effect of the imbalanced distribution of the data between these three datasets (Figure 1).

(A) **Predicted class**

True class	1	2	3
1	0	2.21	2.21
2	1.86	0	1.86
3	109.28	109.28	0

(B) **Predicted class**

True class	1	2	3
1	0	4.59	4.59
2	1.30	0	1.30
3	73.02	73.02	0

Figure 1. Penalty matrices were used for datasets **(A)** (left) and **(B)** (right) based on Equation (1). The rows represent the true classes, while the columns represent the predicted classes.

Various steps had to be completed before using the experimental data for the identification of the chicken behaviors. An overview of the main steps is shown in Figure 2. Important pre-processing steps were video annotation and linking the video data to the IMU sensor data. Raw data from the sensors were pre-processed, for which the Xsens file type changes so that it can be read by data software such as MATLAB. Time windowing was required to generalize data points and extract more information from the acceleration data. Video material was visually annotated once by trained personnel every 0.5 s. Based on the majority of types of behavior in a certain class (see Table 2), an intensity label 1, 2, or 3 scores per 0.5 s time interval. After pre-processing the IMU and annotated video data, time windows containing data of several seconds were created to combine information of multiple datapoints within that specific time window [16,17]. Two different time windows were used, namely a one-second time window and a 4 s time window. The sampling frequency of the IMU was set at 100 Hz, resulting in 100 and 400 IMU data points for the 1 and 4 s time window, respectively. Time windows were shifted 0.5 s which led to a 50% and 87.5% overlap between consecutive time windows of 1 and 4 s, respectively.

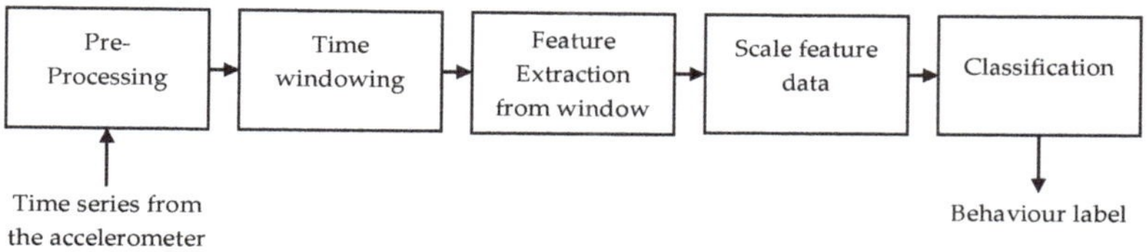

Figure 2. Overview of consecutive steps of the behavior classification process (adopted and modified from Figure 2 in ref [20]).

From the specific data in the windows, it was possible to calculate important characteristics, also called features [21–24]. These features could then be used for the classification of the behavior. A total of 31 features were obtained from the raw accelerometer data. The following features were extracted in X-, Y-, and Z-direction: skewness, kurtosis, mean, standard deviation, variance, minimum, maximum, entropy, energy, and covariance. In addition to these directional features, the average signal magnitude also served as a feature. Feature extraction was performed to reduce dimensionality so that the data could be classified. However, before using the feature data, the data were standardized.

Model creation and a statistical analysis were performed using MATLAB R2020b, version 1.0.0.1. (MathWorks, Inc., Natick, MA, USA) and Microsoft Office Excel 2016. The Random Forest classification model was created in MATLAB R2020b software. The Classification Learner application supported by MATLAB R2020b was used to create a basic Random Forest classification model. The Machine Learning ToolboxTM was used to analyze and model the data using different Machine Learning methods. It provided principal component analysis (PCA), regularization, dimensionality reduction, and feature selection methods that allowed identifying of features with the best predictive power. A two-tailed paired samples *t*-test was used to statistically show a significant difference between results. This was implemented to show whether, for example, a certain difference in an accuracy value would be significantly different or not.

Having a large number of features in a dataset is computationally expensive. In order to reduce the computational load, dimension-reduction techniques with a principal component analysis were applied. Principal component analysis (PCA) is a statistical method that reduces the dimensionality of the dataset by trying to find a low-dimensional representation that captures as much information as possible [25]. This technique assumes that a high variation in data is important for the model. It reduces the number of features, which is called the principal component. This research used PCA to explain at least 95% of the variance of the available data and only included features that contributed to this 95%. The first principal component is the component that explains the largest amount of variance in the data, followed by the principal component that explains the second-largest amount of variance and so on, until the desired 95% of the variance in the dataset was explained.

These data were then exposed to feature extraction, which means reducing the dimensionality, normalizing, and standardizing the dataset. The dataset was then split into a training data set and a test data set, where the training data was a subset used to train a model and the test set was used to test the trained model. Finally, the test data were used to compute the generalization performance of the model, in other words, the ability of the model to generalize the outcome.

2.3. Model Performance Validation

In order to assess the performance of the model, cross-validation and a principal component analysis were introduced. When training the model, overfitting and underfitting were identified and prevented. Cross-validation is a technique that splits the data in a certain way to find the best algorithm for the model. It is used to evaluate the performance of a machine learning model by predicting new datasets that the model has not previously been trained with [26]. This was performed by partitioning the available dataset, using a subset of the whole dataset to train the algorithm and the remaining data of the dataset to test the model so as to evaluate its performance. Each round involved randomly partitioning the original dataset into a training and a test set. This process was then repeated several times, as can be seen in Figure 3.

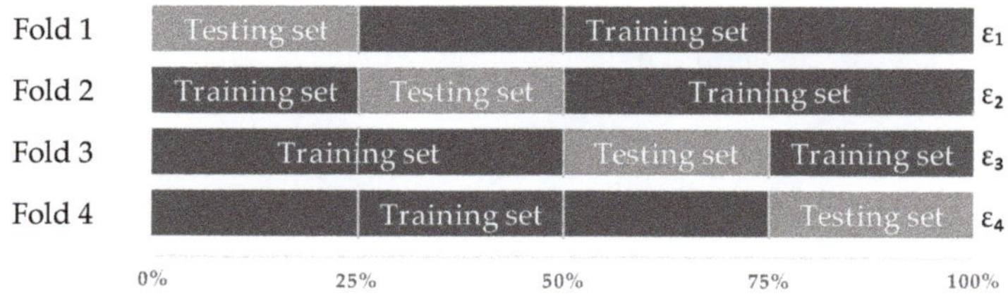

Figure 3. Four-fold Cross-Validation. The original dataset is partitioned into 4 equally sized subsets, which are repeatedly used as either test or training sets.

A confusion matrix was used as a validation tool. The accuracy and F1-scores were calculated to assess model performance [27,28]. The precision and recall were calculated according to:

$$Precision = TP/(TP + FP), \tag{2}$$

$$Recall = TP/(TP + FN), \tag{3}$$

where *TP* is the number of true positives, so where the model correctly predicts the positive class. *FP* is the number of false positives, where the model incorrectly predicts a positive class. *FN* is a false negative, where the model incorrectly predicts the negative class.

Within this research, a multiclass setting instead of a binary classification setting was used. The resulting accuracy and the F1-scores per class were therefore calculated according to:

$$Accuracy = TP/(TP + TN + FP + FN), \tag{4}$$

$$F1\text{-}score\ per\ class = (2 \times Precision \times Recall)/(Precision + Recall), \tag{5}$$

where *TP*, *FP*, and *FN* are the predictions as described above. *TN* is the outcome where the model correctly predicts the negative class. The overall F1-score was then calculated by taking the mean of the individual F1-scores of each class. Figure 4 illustrates a multiclass confusion matrix for a dataset.

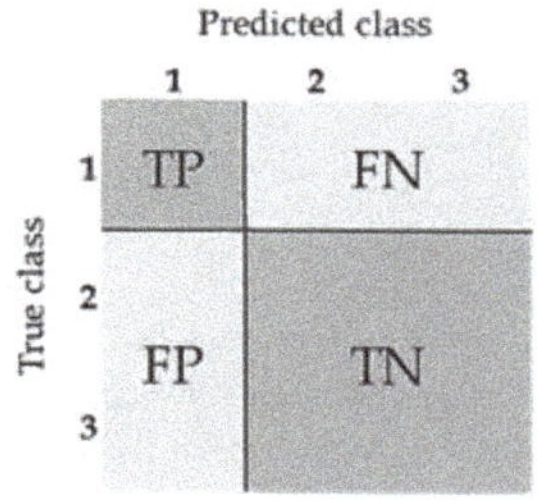

Figure 4. Confusion matrix of true (T), positive (P), false (F), and negative (N) labels for a dataset within a multiclass problem.

3. Results

3.1. Correlation

The variables that are highly correlated with at least one other variable are the standard deviation, the variance, the minimum and maximum, and the energy for each direction (X-, Y-, and Z-direction, Pearson Correlation Coefficient > 0.7 or <−0.7).

For the 4 s time window, there was no statistical difference found between the accuracy values when all variables were included and when some highly correlating variables were removed ($p = 0.987$). However, for the 1 s time window, removing some highly correlating variables decreased the mean accuracy of all the models by 0.5% ($p = 0.0008$).

3.2. Principal Component Analysis

The Principal Component Analysis (PCA) was involved in the next analysis to check its effects on the performance of the different machine learning techniques. With each comparison, the classification models were trained and tested on the same datasets with the same time window, being either 1 s (short) or 4 s (long).

The effect of the randomization on cross-validation (due to randomly assigned subsets) was investigated before the PCA results were analyzed. This was performed by running the models multiple times with the same settings applied. When looking at the 1 s time window with dataset A, there was no significant difference in the accuracy values for both PCA and no PCA ($p = 0.318$ and $p = 0.277$, respectively). The 4 s time window also showed that the randomization due to applying cross-validation did not introduce significant differences with p-values of 0.551 and 0.341 for PCA and no PCA, respectively.

3.3. Time Windows

The F1-score of class 3 was not significantly different. The results are displayed in Table 4. When looking at the 4 s time window and comparing PCA vs. no PCA, not applying PCA on the original dataset resulted in a 1.5% higher accuracy value ($p = 0.035$). The overall F1-score was 2% higher ($p = 0.0008$) when no PCA was applied and the F1-scores of classes 2 and 3 were 1% and 4% higher in the case of no PCA ($p = 0.0108$ and $p = 0.0085$, respectively). The F1-score of class 1 showed no significant difference between PCA and no PCA ($p = 0.175$). Likewise, dataset A showed a 1.3% higher accuracy value when PCA was not applied ($p = 0.0004$). The F1-scores of classes 1 and 2 were both 1% higher resulting in a 3% higher overall F1-score of 0.58 ($p = 0.0185$). The F1-score of classes 1 and 2 were both 0.89 ($p = 0.0113$ and $p = 0.0006$).

Table 4. Overview of the results for the 4 s time window for the original dataset and dataset A. Here, a comparison was made between applying Principal Component Analysis (PCA) or not applying PCA (No PCA). The p-value is provided to show whether the two values are significantly different (p-value $< \alpha$, p-value < 0.05) or not. Significance was tested with a paired samples t-test with a two-tailed distribution.

Parameter	Original Dataset			Dataset A		
	PCA	No PCA	p-Value	PCA	No PCA	p-Value
Accuracy	87.9	89.4	0.0358	87.5	88.8	4.32×10^{-4}
F1-score class 1	0.88	0.89	0.175	0.88	0.89	0.0113
F1-score class 2	0.88	0.89	0.0108	0.88	0.89	5.54×10^{-4}
F1-score class 3	0.42	0.46	8.52×10^{-3}	0.55	0.58	0.214
Overall F1-score	0.73	0.75	7.66×10^{-4}	0.77	0.79	0.0185

The 1 s time window showed significant differences in all measured parameter values for both the original dataset and dataset A except for the F1-score of class 3 and the training time (Table 5). The original dataset had 2.7% higher accuracy values when no PCA was applied compared to the original dataset with PCA applied ($p = 8.5 \times 10^{-8}$). Moreover, the F1-score for class 1 was 2% higher with no PCA applied ($p = 1.43 \times 10^{-9}$). The F1-score for class 2 was also 3% higher in the case of no PCA ($p = 2.32 \times 10^{-6}$). This resulted in a 3% higher overall F1-score of 0.81 when no PCA was applied ($p = 0.0017$). Dataset A showed that not applying PCA resulted in a 2.7% higher accuracy of 83.6% versus an accuracy of 80.9% with PCA enabled ($p = 1.49 \times 10^{-8}$). There were also significant differences in the F1-scores of classes 1 and 2 and the overall F1-score. The F1-scores for classes 1 and 2 were 2% and 3% higher with no PCA, resulting in a 3% higher overall F1-score ($p = 9.86 \times 10^{-9}$,

$p = 2.20 \times 10^{-6}$, and $p = 0.0003$, respectively). The F1-score was not significantly different as the p-value was equal to 0.130.

Table 5. Overview of the results for the 1 s time window for the original dataset and dataset A. Here, a comparison was made between applying Principal Component Analysis (PCA) or not applying Principal Component Analysis (No PCA). The p-value is provided to indicate whether the two values are significantly different (p-value < α, p-value < 0.05) or not. Significance was tested with a paired samples t-test with a two-tailed distribution.

Parameter	Original Dataset			Dataset A		
	PCA	No PCA	p-Value	PCA	No PCA	p-Value
Accuracy	82.8	85.5	8.52×10^{-8}	80.9	83.6	1.49×10^{-8}
F1-score class 1	0.83	0.85	1.40×10^{-9}	0.83	0.85	9.90×10^{-7}
F1-score class 2	0.83	0.86	2.30×10^{-6}	0.81	0.84	2.20×10^{-6}
F1-score class 3	0.67	0.72	0.0780	0.34	0.37	0.130
Overall F1-score	0.78	0.81	1.67×10^{-3}	0.66	0.69	3.10×10^{-4}

For the original dataset with no PCA applied, the overall accuracy is larger for the 4 s time window with an accuracy of 89.4%, compared to an accuracy of 85.5% for the 1 s time window ($p = 0.0028$). The F1-score of class 1 was 4% higher for the 4 s time window and the F1-score of class 2 was 3% higher ($p = 5.58 \times 10^{-5}$ and $p = 1.3 \times 10^{-4}$). The 4 s time window had a lower F1-score for class 3 with the F1-score being 0.46 versus 0.72 for the 1 s time window ($p = 1.04 \times 10^{-7}$). Subsequently, the overall F1-score was significantly higher for the 1 s time window with an overall F1-score of 0.80 versus an overall F1-score of 0.75 for the 4 s time window ($p = 8.63 \times 10^{-5}$).

For dataset A with no PCA applied, there were differences in all of the measured outcomes, except for the training time (Table 6). The models had an average accuracy of 88.8% for the 4 s time window and 83.6% for the 1 s time window ($p = 5.48 \times 10^{-5}$). The F1-scores for the 4 s time window were 0.89, 0.89, and 0.58, respectively, for classes 1, 2, and 3, resulting in an overall F1-score of 0.79. The 1 s time window resulted in F1-scores of 0.85, 0.82, and 0.37 for, respectively, for classes 1, 2, and 3. This resulted in an overall F1-score of 0.68. These values were statistically different with the p-values being 0.013, 2.67×10^{-6}, 4.07×10^{-7}, and 9.25×10^{-8}, respectively, for the F1-score of class 1, 2, and 3, and the overall F1-score. Since the type of highly dynamic behavior is very short and a sharp peak in the acceleration data is often observed, the time window used is important. When time windows are longer, they contain more data of that specific event. This increases the amount of information of that event, which should increase the classification performance of the model. The results are displayed in Table 6.

Table 6. Overview of the results for the long time window (4 s, called Long) and the short time window (1 s, called Short) for datasets original and A. No PCA was applied. The p-value is provided to indicate whether the two values are significantly different (p-value < α, p-value < 0.05) or not. Significance was tested with a paired samples t-test with a two-tailed distribution.

Parameter	Original Dataset			Dataset A		
	Long	Short	p-Value	Long	Short	p-Value
Accuracy	89.4	85.5	2.82×10^{-4}	88.8	83.6	5.48×10^{-5}
F1-score class 1	0.89	0.85	5.58×10^{-5}	0.89	0.85	1.31×10^{-2}
F1-score class 2	0.89	0.86	1.31×10^{-4}	0.89	0.84	2.67×10^{-6}
F1-score class 3	0.46	0.72	1.04×10^{-7}	0.58	0.37	4.07×10^{-7}
Overall F1-score	0.75	0.81	8.63×10^{-5}	0.79	0.69	9.25×10^{-8}

The more precisely annotated labels for class 3 resulted in different performances for the different time windows with dataset A. The machine learning methods showed an increase in the F1-score of class 3 for the 4 s time window when extra data for this class was provided. However, on the 1 s time window, the F1-score showed a different result, with a lower F1-score for class 3 when extra data for this class was provided. The 4 s time window with no PCA applied showed an increase in the F1-score of class 3 from 0.46 in the original dataset to 0.58 in dataset A ($p = 3.00 \times 10^{-7}$). This led to an increase in the overall F1-score from 0.75 to 0.79 for, respectively, the original dataset and dataset A ($p = 2.00 \times 10^{-6}$). The other measured parameters such as accuracy and the F1-scores for classes 1 and 2 did not show significant differences. The 1 s time window showed a decrease of the F1-score of class 3 with 0.72 for the original dataset and 0.37 for dataset A ($p = 1.68 \times 10^{-9}$). This reduced the overall F1-score from 0.81 to 0.69 ($p = 1.89 \times 10^{-9}$).

3.4. Machine Learning Models

A comparison was made between an existing Random Forest model and the other available models to find the best model for this study. This means that the original dataset with a 1 s time window was selected without applying PCA. When comparing these results, the level of significance is important in order to indicate whether the results are significantly different. Therefore, two situations were investigated: one situation with the default significance level of 5% ($p < 0.05$) and one situation with a significance level of 10% ($p < 0.1$).

When looking at the default significance level, we found that there were no models significantly better than the Random Forest model. When altering the significance level to a level that allows less than 1 in 10 chance of being wrong (p-value of 0.1), two models showed significantly higher values. These models were the Bagged Trees and subspace KNN within the ensemble Machine Learning technique. Their p-values were, respectively, 0.054 and 0.097 and provided a significantly better fit on the data than the initial Random Forest model. In Random Forest, only a subset of features is selected at random, while with Bagged Trees all features are considered when splitting a node. In this research, Bagged Trees performed best. Therefore, only the results of the Bagged Trees model will be further considered.

The Bagged Trees model had an accuracy value of 90.0% and its F1-scores were 0.89, 0.91, and 0.87 for, respectively, classes 1, 2, and 3. This resulted in an overall F1-score of 0.89. Compared to the Random Forest model the accuracy was increased by 1% and the increase in F1-scores showed a better fit on the different classes. The F1-scores increased by 2%, 1%, and 6% for classes 1, 2, and 3, resulting in an overall F1-score increase of 3%. The confusion matrices of the Random Forest model and the Bagged Trees model are provided in Figure 5.

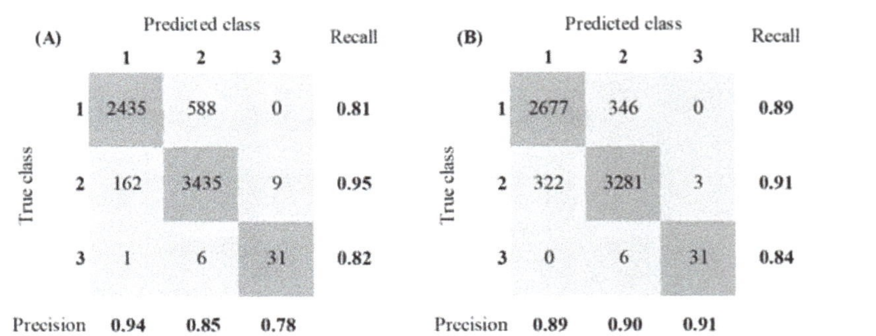

Figure 5. Confusion matrices of the Random Forest model (**A**) and the Bagged Trees model (**B**). (Class 1 = static behavior, class 2 = semi-dynamic behavior, class 3 = high-dynamic behavior).

3.5. Model Validation Based on the Second Chicken (Dataset B)

To verify the classification performance of the Bagged Trees model, this model was exposed to accelerometer data and labelled data of a different chicken (blue chicken, dataset B). The confusion matrices of both models with dataset B as input data are given in Figure 6.

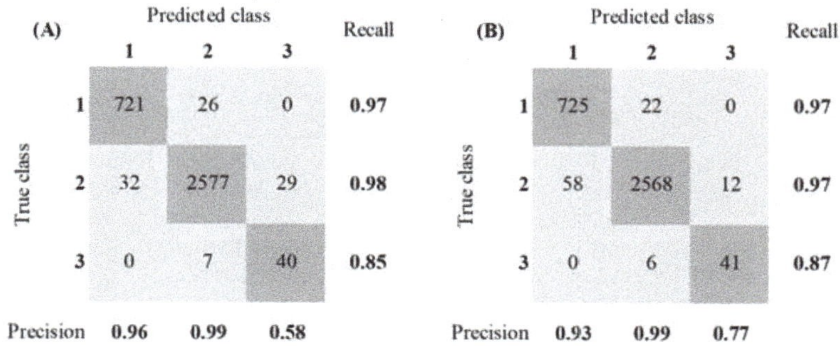

Figure 6. Confusion matrix of the Random Forest model (**A**) and the Bagged Trees model (**B**) with dataset B as input data and no PCA applied. (Class 1 = static behavior, class 2 = semi-dynamic behavior, class 3 = high-dynamic behavior).

The same data-handling method was used to check the classification performance (short time window and no PCA applied). The Bagged Trees model was able to predict the data with an accuracy of 97.1%. This resulted in F1-scores of 0.95, 0.98, and 0.82 of classes 1, 2, and 3. The overall F1-score was equal to 0.92. Comparing these results with the results of the Random Forest model showed that the F1-score of class 1 decreased by 1% when using the Bagged Trees model instead of the Random Forest model. The F1-score of class 2 remained equal. Nonetheless, the Bagged Trees model was able to increase the F1-score of class 3 by 13%, increasing from 0.69 for the Random Forest model to 0.82 by using the Bagged Trees model. Subsequently, the overall F1-score increased by 4% (0.88 for the Random Forest model and 0.92 for the Bagged Trees model).

4. Discussion

4.1. Data Collection

Data were collected from two laying hens, from the same genotype and housed in the same environment over a limited amount of time. This could have influenced the results as other genotypes may have slightly different behavioural patterns and some environments allow more behavioural expressions than others. For this study, however, the limitations on the results are not thought to have a considerable effect. The housing system of the hens allowed them to express a normal variety of behaviors. Different genotypes, more hens and longer data recording could influence the distribution of the individual behaviors, but the classification of behaviors as used in this study will fade out small differences between breeds or caused by differences in, e.g., climate [29]. As long as these differences do not cause a completely different number of data points per behavioural class, no substantial effect is expected on the performance of the models. Therefore, the results are expected to be applicable for other housing systems and other genotypes. For application in large flocks, additional research is required to identify the minimum or the optimum number of hens that need to be equipped with sensors for reliable behavioural assessment in various parts of the henhouse.

Monitoring more than two chickens for more days might lead to more data to predict the underlying behavior of laying hens. However, because of animal-welfare concerns, it was not possible to run experiments for more than 5 days. Additionally, more data are acquired across more days, but these data will not necessarily provide information regarding the chicken behavior, therefore, the accuracy of the model will not significantly improve. In other words, it is necessary to acquire more data if one aims to identify specific behaviours of chickens while the behaviour of chickens was classified into three classes in this research and the number of chickens and the measurement days were sufficient enough for this aim. Considering the fact that the inertia sensors are relatively expensive equipment, using only two chickens to identify the chicken behaviors is one of the strengths of this research study.

4.2. Principal Component Analysis (PCA)

Removing highly correlating variables from the training procedure did not result in a significant improvement to model performance. This was tested to see whether the behavior of one variable could be predicted by the behavior of the other correlating variable. Initially, the non-correlated variables seemed to be independent and could help to recognize the activity level within the datasets. The only reason to remove highly correlated features would be for storage and computational load, but as these were not a problem, all variables were kept within the models. Moreover, removing the highly correlated variables resulted in decreased model accuracies. The use of Principal Component Analysis was investigated since it automatically selects a set of orthogonal Principal Components (non-correlating variables) to train the models. Application of Principal Component Analysis (PCA) did not introduce significant improvements to the model outputs. In most cases, PCA reduced the capacity of the models to correctly classify laying-hen behavior [30].

4.3. Cross-Validation

Besides the PCA, cross-validation can also be used as a way to prevent overfitting [31]. Since all the datasets were first exposed to cross-validation, the datasets were already evaluated based on their capacity to generalize. As the models were cross-validated using 4-fold cross-validation, the accuracy values of some scenarios were different when the models ran multiple times. As this did not show a significant difference, only a single run of all the models was considered to be part of the results. To decrease the effect of randomization during cross-validation even more, future research could run the models multiple times instead of only a single run and combine the results and take, e.g., the mean value for the accuracy. This should decrease the effect of the randomization part during cross-validation and will, therefore, increase the validity of the results of the models.

This research has shown that machine learning is able to detect laying hen behavior under different circumstances. The accuracy of the models was in all different scenarios above 70%.

As mentioned before, MATLAB used cross-validation to prevent overfitting. The selected method for this procedure was k-fold cross-validation and this includes randomization, as k-fold cross-validation partitions the data into k randomly assigned subsets of equal size [32]. This randomization caused the models to have slightly different results in some instances. In order to check the effect of the randomization on the final output of the models, the models have been run several times with the same settings applied. The results of this effect did not have a significant impact on the accuracy values of the models, so the accuracy values of one-run were considered in each scenario.

Additionally, this research focused more on the activity level of laying hens rather than the specific behaviors of laying hens. This way of classifying laying hen behavior is too general to obtain an accurate prediction of the precise laying hen behavior. Such a general prediction of laying hen behavior may be less useful for behavioural studies [18], but it may provide a useful application in commercial poultry, as commercial farmers are often more interested in flock behavior and general activity levels, rather than single specific behaviors. As a consequence, the models in this research can be beneficial in the development of precision livestock farming in the poultry sector [33].

4.4. Highly Dynamic Behavior

The obtained results for the high-dynamic class, class 3, show that the machine learning models were able to accurately capture this type of behavior. Due to the fact that the type of highly dynamic behavior is very short and often presents a sharp peak in the acceleration data, the time window used is important [34]. When time windows are longer, they contain more information, which increases the classification performance of the model. This was also the case when the results of the 4-s time window were compared with the 1-s time window. The 4-s time window had significantly higher accuracy values than the 1-s time window. Additionally, the F1-scores of classes 1 and 2 are in most instances significantly higher for the 4-s time window. This is due to a larger number of data points for these two classes. However, the F1-score of class 3 was significantly better when using the shorter time window. Model prediction was harder as there were fewer data points of class 3 compared to classes 1 and 2. As fewer data provide the model with fewer data to make correct predictions. Due to the low amount of data points of class 3, a 4 s time window will generally have fewer assigned data points of this class compared to a 1 s time window. This explains the difficulty of correctly predicting class 3 with a 4 s time window [14,34].

When highly dynamic behavior is of greater interest than static behavior, a shorter time window will lead to higher F1-scores and is, therefore, more favourable than a longer time window. However, using a shorter time window is at the expense of having lower classification performance on static- and semi-dynamic behavior.

As shown by Calvet et al. [9], there is a relationship between the behavior of chickens and dust concentration inside the indoor air, as also observed by Winkel [4]. The obtained results of this research indicated that the highly dynamic behaviors such as scratching/dust bathing can be determined with acceptable accuracy even with a few data sets. Distinguishing the chicken behavior can help to identify the distribution of different dynamic behavior in time or its spread out over the day. Consequently, this information can be used in air-quality control and reduction of indoor dust concentration and fine dust emission into the environment (e.g., temporarily lower/higher ventilation rates, or applying fine-dust-reduction techniques more intensively) [11].

4.5. Performance of the ML Models

The largest improvement of using the Bagged Trees model over a Random Forest model was found in the classification performance of highly dynamic behaviors. Both the Bagged Trees model and the Random Forest model draw random bootstrap samples from the training set. However, besides the bootstrap samples, the Random Forest model draws random subsets of features for training the individual trees, while in bagging the full set of features is provided to each tree. This random feature selection in the Random Forest causes the trees to become more independent of each other compared to regular bagging. This research has shown that this random feature selection causes a worse classification performance for highly dynamic behaviors, while not significantly impacting the performance of the model on static- and semi-dynamic behaviors. Therefore, the regular bagged tree model is preferred over the Random Forest model [35].

4.6. Sensor Technology

There are multiple inertial sensors available. The main distinction is made between high-end inertial sensors and low-cost inertia sensors or custom-made inertial sensors. Two types of high-end inertial sensors are the Xsens IMU, which was used in this study, and the Physilog 5. These sensors distinguish themselves from the low-cost sensors due to their ability to measure the 3D acceleration, the 3D rate of turn, and the 3D magnetic field all in one device. Additionally, data are easily transferrable with USB, and they are water and dust resistant with IP67/IP64 certification.

Some examples of low-cost inertial sensors are the Sparkfun, the WitMotion, and Zstar3. The limitations of low-cost sensors are often that they suffer from poor signal to noise ratios and limited dynamic ranges. Nandy et al. [36] provided a way to produce a custom-made inertial sensor.

5. Conclusions

The novelty of this study was that it proved the feasibility of using machine learning models and inertia sensors to identify and classify laying-hen activity at various levels. We showed a proof of principle of tools for analyzing laying-hen behaviors by using wearable inertia sensor technology and machine learning models (ML).

The machine learning models of this study can predict the three activity levels of laying hens with overall accuracy values of over 90%. Removing highly correlating variables did not introduce significant model improvement with the original dataset. Additionally, applying PCA did not result in better model classification performance. Static behaviour and semi-dynamic behaviour can best be analysed with a long time window, whereas highly dynamic behaviour favours a short time window. Annotating the data of class 3 more precisely only increased model performance for the long time window. The best performing machine learning model was the bagged tree model with an accuracy of 90% for the original dataset.

The methodology developed in this paper can be used in the development of precision livestock farming systems for the commercial poultry sector. The outcomes of this research show that machine learning is able to accurately analyse different activity levels of laying-hen behaviour. This provides a reason for further research to analyse laying-hen behaviour in more detail in the future.

Author Contributions: S.M.D. designed and directed the project; T.G.C.M.v.N. co-supervised the project; P.W.G.G.K. co-supervised the project; M.O. developed the models and analysed the data; All authors provided critical feedback and helped shape the research, analysis the results, and writing of the manuscript. All authors have read and agreed to the published version of the manuscript.

Funding: This research received no external funding.

Institutional Review Board Statement: Ethical approval for the experiment was granted by the Animal Welfare Body of Wageningen Research (date: 8 November 2019), for mounting backpacks on two chickens for a maximum period of 5 days, which was not exceeded.

Data Availability Statement: The data presented in this study are available on request from the corresponding author. The data are not publicly available because they contain information that could compromise the privacy of the farmer.

Conflicts of Interest: The authors declare no conflict of interest.

References

1. Council of the European Union. Council Directive 99/74/EC of 19 July 1999 laying down minimum standards for the protection of laying hens. *Off. J. Eur. Communities* **1999**, *203*, 53–57.
2. EFSA. Opinion of the Scientific Panel on Animal Health and Welfare (AHAW) on a request from the Commission related to the welfare aspects of various systems of keeping laying hens. *EFSA J.* **2005**, *3*, 197. [CrossRef]
3. Elson, H.A. Poultry welfare in intensive and extensive production systems. *Worlds Poult. Sci. J.* **2015**, *71*, 449–460. [CrossRef]
4. Winkel, A. Particulate Matter Emission from Livestock Houses: Measurement methods, Emission Levels and Abatement Systems. Ph.D. Thesis, Wageningen University & Research, Wageningen, The Netherlands, 21 October 2016. Available online: https://library.wur.nl/WebQuery/wurpubs/508819 (accessed on 11 January 2022).
5. Casey, K.; Bicudo, J.; Schmidt, D.; Singh, A.; Gay, S.; Gates, R.; Jacobson, L.; Hoff, S. Air Quality and Emissions from Livestock and Poultry Production/Waste Management Systems. In *Animal Agriculture and the Environment: National Center for Manure and Animal Waste Management White Papers*; American Society of Agricultural and Biological Engineers: Saint Joseph, MI, USA, 2020.
6. Takai, H.; Pedersen, S.; Johnsen, J.O.; Metz, J.H.M.; Groot Koerkamp, P.W.G.; Uenk, G.H.; Phillips, V.R.; Holden, M.R.; Sneath, R.W.; Short, J.L.; et al. Concentrations and Emissions of Airborne Dust in Livestock Buildings in Northern Europe. *J. Agric. Eng. Res.* **1998**, *70*, 59–77. [CrossRef]
7. Cambra-López, M.; Aarnink, A.J.A.; Zhao, Y.; Calvet, S.; Torres, A.G. Airborne particulate matter from livestock production systems: A review of an air pollution problem. *Environ. Pollut.* **2010**, *158*, 1–17. [CrossRef]
8. Directive Council. Directive 2010/75/EU of the European Parliament and of the Council. *Off. J. Eur. Union L* **2010**, *334*, 17–119.
9. Calvet, S.; Van den Weghe, H.; Kosch, R.; Estellés, F. The influence of the lighting program on broiler activity and dust production. *Poult. Sci.* **2009**, *88*, 2504–2511. [CrossRef]
10. Ellen, H.H.; Bottcher, R.W.; von Wachenfelt, E.; Takai, H. Dust Levels and Control Methods in Poultry Houses. *J. Agric. Saf. Health* **2000**, *6*, 275–282. [CrossRef]
11. Van Niekerk, T. Managing laying hen flocks with intact beaks. In *Achieving Sustainable Production of Eggs Volume 2*; Burleigh Dodds Science Publishing: Cambridge, UK, 2017; pp. 163–176, ISBN 9781351114134.
12. Dawkins, M.S. Behaviour as a tool in the assessment of animal welfare. *Zoology* **2003**, *106*, 383–387. [CrossRef]

13. Herbert, G.T.; Redfearn, W.D.; Brass, E.; Dalton, H.A.; Gill, R.; Brass, D.; Smith, C.; Rayner, A.C.; Asher, L. Extreme crowding in laying hens during a recurrent smothering outbreak. *Vet. Rec.* **2021**, *188*, e245. [CrossRef]
14. Kozak, M.; Tobalske, B.; Springthorpe, D.; Szkotnicki, B.; Harlander-Matauschek, A. Development of physical activity levels in laying hens in three-dimensional aviaries. *Appl. Anim. Behav. Sci.* **2016**, *185*, 66–72. [CrossRef]
15. Hansen, I. Behavioural expression of laying hens in aviaries and cages: Frequencies, time budgets and facility utilisation. *Br. Poult. Sci.* **1994**, *35*, 491–508. [CrossRef]
16. Hepworth, P.J.; Nefedov, A.V.; Muchnik, I.B.; Morgan, K.L. Broiler chickens can benefit from machine learning: Support vector machine analysis of observational epidemiological data. *J. R. Soc. Interface* **2012**, *9*, 1934–1942. [CrossRef] [PubMed]
17. Oladipupo, T. Types of Machine Learning Algorithms. In *New Advances in Machine Learning*; InTech: Vienna, Austria, 2010.
18. Stadig, L.M.; Rodenburg, T.B.; Ampe, B.; Reubens, B.; Tuyttens, F.A.M. An automated positioning system for monitoring chickens' location: Effects of wearing a backpack on behaviour, leg health and production. *Appl. Anim. Behav. Sci.* **2018**, *198*, 83–88. [CrossRef]
19. Tom, R. Monitoring and Analysis of the Daily Behaviour of Individual Laying. Ph.D. Thesis, Wageningen University & Research, Wageningen, The Netherlands, 8 April 2020.
20. Rahman, A.; Smith, D.V.; Little, B.; Ingham, A.B.; Greenwood, P.L.; Bishop-Hurley, G.J. Cattle behaviour classification from collar, halter, and ear tag sensors. *Inf. Process. Agric.* **2018**, *5*, 124–133. [CrossRef]
21. Robert, B.; White, B.J.; Renter, D.G.; Larson, R.L. Evaluation of three-dimensional accelerometers to monitor and classify behavior patterns in cattle. *Comput. Electron. Agric.* **2009**, *67*, 80–84. [CrossRef]
22. Wang, Y.; Nickel, B.; Rutishauser, M.; Bryce, C.M.; Williams, T.M.; Elkaim, G.; Wilmers, C.C. Movement, resting, and attack behaviors of wild pumas are revealed by tri-axial accelerometer measurements. *Mov. Ecol.* **2015**, *3*, 2. [CrossRef]
23. Twomey, N.; Diethe, T.; Fafoutis, X.; Elsts, A.; McConville, R.; Flach, P.; Craddock, I. A Comprehensive Study of Activity Recognition Using Accelerometers. *Informatics* **2018**, *5*, 27. [CrossRef]
24. Peng, Z.; Zhang, Y. Dilemma and Solution of Traditional Feature Extraction Methods Based on Inertial Sensors. *Mob. Inf. Syst.* **2018**, *2018*, 2659142. [CrossRef]
25. Picasso, P. *Principal Component Analysis*; Everitt, B.S., Howell, D.C., Eds.; Springer Series in Statistics; Springer: New York, NY, USA, 2002; Volume 9, ISBN 0-387-95442-2.
26. Bergmeir, C.; Benítez, J.M. On the use of cross-validation for time series predictor evaluation. *Inf. Sci.* **2012**, *191*, 192–213. [CrossRef]
27. Zhang, D.; Wang, J.; Zhao, X. Estimating the uncertainty of average F1 scores. In Proceedings of the 2015 International Conference on the Theory of information Retrieval, Northampton, MA, USA, 27–30 September 2015; pp. 317–320. [CrossRef]
28. Chicco, D.; Jurman, G. The advantages of the Matthews correlation coefficient (MCC) over F1 score and accuracy in binary classification evaluation. *BMC Genom.* **2020**, *21*, 6. [CrossRef] [PubMed]
29. Ellen, E.; van der Sluis, M.; Siegford, J.; Guzhva, O.; Toscano, M.; Bennewitz, J.; van der Zande, L.; van der Eijk, J.; de Haas, E.; Norton, T.; et al. Review of Sensor Technologies in Animal Breeding: Phenotyping Behaviors of Laying Hens to Select Against Feather Pecking. *Animals* **2019**, *9*, 108. [CrossRef] [PubMed]
30. Jolliffe, I.T.; Cadima, J. Principal component analysis: A review and recent developments. *Philos. Trans. R. Soc. A Math. Phys. Eng. Sci.* **2016**, *374*, 20150202. [CrossRef] [PubMed]
31. Berrar, D. Cross-Validation. In *Encyclopedia of Bioinformatics and Computational Biology*; Elsevier: Amsterdam, The Netherlands, 2019; pp. 542–545.
32. Rodriguez, J.D.; Perez, A.; Lozano, J.A. Sensitivity Analysis of k-Fold Cross Validation in Prediction Error Estimation. *IEEE Trans. Pattern Anal. Mach. Intell.* **2010**, *32*, 569–575. [CrossRef] [PubMed]
33. Sassi, N.B.; Averós, X.; Estevez, I. Technology and Poultry Welfare. *Animals* **2016**, *6*, 62. [CrossRef]
34. Banerjee, D.; Biswas, S.; Daigle, C.; Siegford, J.M. Remote Activity Classification of Hens Using Wireless Body Mounted Sensors. In Proceedings of the 2012 Ninth International Conference on Wearable and Implantable Body Sensor Networks, London, UK, 9–12 May 2012; IEEE: New York, NY, USA, 2012; pp. 107–112.
35. Bannerman-Thompson, H.; Bhaskara Rao, M.; Kasala, S. Bagging, boosting, and random forests using R. In *Handbook of Statistics*; Elsevier: Amsterdam, The Netherlands, 2013; Volume 31, pp. 101–149.
36. Nandy, A.; Chakraborty, S.; Chakraborty, J.; Venture, G. Low-cost sensors for gait analysis. *Mod. Methods Afford. Clin. Gait Anal.* **2021**, 25–44. [CrossRef]

animals

MDPI

Article

Assessment of Husbandry Practices That Can Reduce the Negative Effects of Exposure to Low Ammonia Concentrations in Broiler Houses

Leonardo V. S. Barbosa [1,*], Daniella J. De Moura [1,*], Fernando Estellés [2], Adrian Ramón-Moragues [3], Salvador Calvet [2] and Arantxa Villagrá [3]

1 College of Agricultural Engineering, State University of Campinas, 501 Candido Rondon Avenue, São Paulo 13083-875, Brazil
2 Institute of Animal Science and Technology, Universitat Politècnica de València, Camino de Vera s.n., 46022 Valencia, Spain; feresbar@upv.es (F.E.); salcalsa@upvnet.upv.es (S.C.)
3 Centro de Investigación en Tecnología Animal (CITA), Valencian Institute for Agricultura Research (IVIA), 12400 Segorbe, Spain; armzulu@gmail.com (A.R.-M.); villagra_ara@gva.es (A.V.)
* Correspondence: valentino.leo@icloud.com (L.V.S.B.); djmoura@unicamp.br (D.J.D.M.); Tel.: +55-19-984417811 (L.V.S.B.)

Simple Summary: We used two commercial breeds, differing in growth rate: Fast-growing breed and slow-growing breed. We stocked these birds in two different densities. The slow-growing birds was stocked at a high density and the fast-growing birds at a high density and low density. These birds were reared under two different environmental conditions: A control room with a low concentration of ammonia and a second room with a higher concentration. We analyzed management practices such as the effect of ventilation, animal density and growth rate as management possibilities to reduce the negative effect of ammonia on production parameters.

Citation: Barbosa, L.V.S.; De Moura, D.J.; Estellés, F.; Ramón-Moragues, A.; Calvet, S.; Villagrá, A. Assessment of Husbandry Practices That Can Reduce the Negative Effects of Exposure to Low Ammonia Concentrations in Broiler Houses. *Animals* **2022**, *12*, 1096. https://doi.org/10.3390/ani12091096

Academic Editor: Marian Stamp Dawkins

Received: 26 February 2022
Accepted: 20 April 2022
Published: 23 April 2022

Abstract: Ammonia is an important pollutant emitted by broiler litter that can accumulate inside farms, impairing animal health and welfare productivity. An experiment was designed to evaluate of precision husbandry practices such as the effect of ventilation, animal density and growth rate as management options to reduce the adverse effects of ammonia exposure on productive parameters in broiler houses. Two identical experimental rooms were used in this study. They were programmed to differ in ammonia concentration from day 32 of the growing period (10 and 20 ppm in Room 1 and Room 2, respectively). Three treatments were tested in each room: slow growth in high stocking density (SHD), fast growth in low density (FLD) and fast growth in high density (FHD). Animal weight, feed intake and feed conversion ratio were determined weekly. In addition, the immune status of animals was assessed by weighing the organs related to immune response as stress indicators. Increasing ventilation was effective to control ammonia concentrations. Exposure to ammonia caused no significant effect on productive parameters. However, lowering stocking density improved response to higher ammonia concentrations by lowering the feed conversion ratio. No other relevant effects of differential exposure to ammonia were found in fast-growing animals, either at high or low stocking density. The use of slow-growing breeds had no effect on production parameters. Despite having a slower growth rate, their feed conversion ratio was not different from that of fast-growing breeds. The productive performance of slow-growing animals was not affected by the differential exposure to ammonia, but the reduced spleen size would suggest an impairment of the immune system.

Keywords: animal welfare; animal health; immune system; productive parameters; management

1. Introduction

Maintaining a proper rearing environment is essential in broiler production. Ammonia concentration in the air is one of the main factors impairing broiler health, welfare and productivity. Exposure of animals to high ammonia levels causes irritation of mucous membranes and the respiratory tract, conjunctivitis and dermatitis [1–4]. Animals exposed to ammonia reduce feed consumption and consequently productivity is impaired through lower growth rate, higher mortality and worse feed conversion ratios [5–8]. It has also been demonstrated that high ammonia concentrations indirectly influence gene expression, affecting the immune response [9] and productivity in terms of slaughter performance and breast yield [10]. For those reasons, welfare regulations establish a concentration threshold for ammonia of 20 ppm according to European Directive 2007/43/CE.

However, the implications of levels below 20 ppm throughout the broiler production cycle are not well known. Most studies do not report relevant effects at concentrations lower than 20 ppm. However, there is evidence that animals' response starts under lower concentrations. It has been described that concentrations of 15 ppm did not affect growth performance, but induced an anti-inflammatory response in the ileum and altered the tracheal microbiota, causing respiratory tract injury [11,12]. Therefore, understanding the effects of concentrations lower than 20 ppm is necessary.

Precision husbandry practices are useful tools to reduce the negative impacts of ammonia exposure. More resilient animals, reducing animal density or increasing ventilation rates are effective practices on farms. Growth rate and animal density are related to animal resilience. Slow-growing broilers are healthier and express more behavioral indicators of positive welfare [13]. Additionally, it has been reported that slow growing animals have a greater magnitude of innate immune response to infections [14]. Therefore, broilers with a lower growth rate may be expected to be more robust against ammonia exposure and less affected in terms of welfare, health and productivity. However, to the authors' knowledge, there is no information on how ammonia exposure affects slow-growing broilers differently from conventional broiler breeds. The effect of stocking density on broiler performance has been described in the literature [15], but the interacting effect of stocking density with the exposure to ammonia is not clear.

Engineering solutions are also available to control ammonia concentrations in broiler farms. The concentration of ammonia in the indoor air is therefore influenced by the diluting effect of ventilation. Ammonia comes from the breakdown of the uric acid excreted by the broilers [16]. The emission rate depends on litter moisture, and litter N content tends to increase as excreta is accumulated in bedding material [17]. Once emitted, the ammonia accumulates inside broiler houses until it is emitted to the atmosphere through the ventilation system. Adjusting ventilation is a strategic mechanism for maintaining an appropriate indoor air quality. It is possible to adjust ventilation to achieve proper indoor air quality, but this normally involves higher energy consumption [18]. Therefore, farmers need to have clear evidence of the welfare and productivity benefits of maintaining ammonia concentrations below certain thresholds.

The hypothesis motivating this study is that there may be additional benefits in reducing stressing factors (ammonia exposure, animal density or growth rate) below the common practice on farm. However, there is scarce information quantifying these effects. Therefore, the objective of this study was to evaluate the influence of two ammonia concentrations (10 and 20 ppm), two broiler breeds (fast vs. medium growth rate) and stocking density (13 vs. 6.5 broiler/m^2) on the productive performance and physiological parameters during one rearing cycle.

2. Materials and Methods

2.1. House Description

The experiment was conducted in accordance with the animal research regulations of the EU, with protocol number 2018/VSC/PEA/0067. The test was carried out at the Animal Technology and Research Center (CITA-IVIA), located in Segorbe, (Castellón, Spain). Two identical rooms (Room 1 and Room 2) were used in this trial. Each room had a dimension of 13.2 m × 5.95 m and had independent mechanical ventilation through ceiling fans (Figure 1).

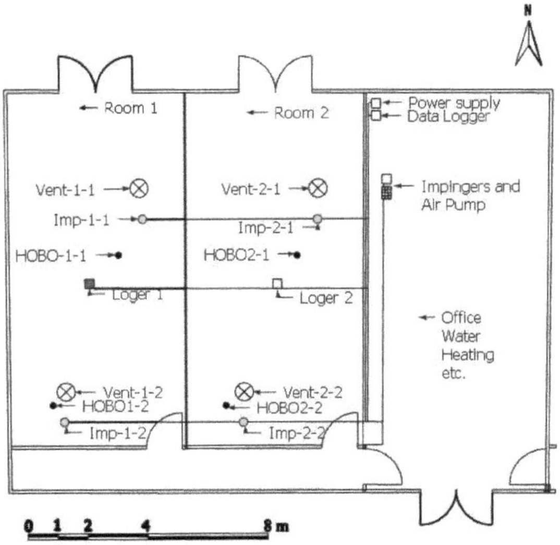

Figure 1. Top view of broiler house. Location of fans (Vent), temperature and relative humidity loggers (HOBO), and ammonia sensors is indicated.

An automated climate control system (DNP Climate Controller, Exafan, Spain) regulated ventilation rates according to commercial breed recommendations of temperature. The environmental temperature was gradually decreased from 32 °C (day 1) to 19 °C (day 42). Temperature was controlled using the climate controller sensor and was recorded together with relative humidity every 10 min using a data logger (HOBO U12, Onsetcomp, Bourne, MA, USA). Additionally, each room was equipped with one electrochemical ammonia sensor (DOL 53, Dräger, Germany). The location of these sensors is presented in Figure 1.

Ammonia was established as a criterion to operate ventilation. In Room 1, ventilation was programmed to maintain a maximum of 10 ppm ammonia, while in Room 2, it was programmed to reach a maximum of 20 ppm. These concentration conditions were programmed to be maintained from the fourth week, that is, in the second half of the production cycle, which is when the ammonia levels are usually higher inside the farms. From the start of the growing period until week 4 of age, both rooms were ventilated following an identical program based on temperature control and animal age. A propane heater was used to maintain adequate room temperatures. Wood shavings (5 cm) were used as the litter material.

The experiment was conducted in winter, when gas concentrations were expected to be higher due to lower ventilation rates. In order to promote NH_3 emissions and achieve the desired concentrations of ammonia, a urea solution was applied to the litter. Urea was applied manually using a backpack sprayer on days 32, 39, 51 and 56 of the growing period at a dose of 0.21 L m^{-2} of urea solution in distilled water.

2.2. Animals and Experimental Design

Two commercial breeds were used, differing in growth rate. The fast-growing breed was Ross®, with a slaughter age of 42 days. The slow-growing breed was Hubbard®, with a slaughter age of 63 days. Three treatments were tested. The first treatment (SHD) was slow-growing birds stocked at a high density (32 kg/m^2 at slaughter age). The second treatment (FHD) was fast-growing birds at a high density (32 kg/m^2 at slaughter age). The third treatment (FLD) was fast-growing birds at a final stocking density of 16 kg/m^2.

The diet was provided ad libitum. Both commercial breeds received a commercial diet. The feed for the fast-growing breed stocked in high and low density consisted of the Nanta A80, and for the slow growing breed, Nanta A32.

In each room, 18 collective pens were installed to allocate 6 repetitions of each treatment. Each pen was provided with 3 nipple drinkers and a manual feeder and housed 17 birds. The dimensions of high-density pens, corresponding to SHD and FHD treatments, were 1×1.3 m^2, whereas low-density pens (corresponding to FLD) had the same number of animals in double the surface area (2×1.3 m^2). Therefore, in each room, 204 fast-growing and 102 slow-growing birds were housed inside the pens. As the pens only occupied a part of each room, 319 broilers were allocated in the space outside the pens in each room. This was completed to simulate the real ambient conditions inside a broiler house, where ammonia comes from litter. Therefore, a total of 625 birds were used in each room. All birds were fed with commercial starter feed from day 1 to day 18 of testing, and from day 18 until the end of the test, each strain was fed with different feeds according to their demands (Figure 2).

Figure 2. Experimentation room with details of the layout of the experimental compartments.

2.3. *Animal Performance and Health*

All the animals housed in the pens were weighed weekly. Feed consumption was also measured in each pen on a weekly basis. The average daily weight gain (DWG) per bird was subsequently calculated for each week of rearing and that accumulated during the test (42 days FLD and FHD and 63 for SHD). The feed conversion rate (FCR) was also calculated for each week and at a cumulative level by dividing the amount of feed consumed by each pen by the growth of all the birds in it.

Thirty animals per treatment were slaughtered on days 21 and 42 to measure the weight of organs related with the immune system. Slow-growing animals were also sampled on day 63. Animals were slaughtered with electric stunning. The immune organs taken from each animal were the spleen, thymus gland, liver and bursa of Fabricius. Organs were weighed using a precision scale (Ohaus Pioneer, PX, Nanikon, Switzerland). Afterwards, weight ratios of each organ were made according to the weight of the birds from which it came to check physiological and anatomical disorders in the immune system organs. On day 63, Fabricio's bag could not be obtained due to its small size or nonexistence, as this organ tends to disappear as animals grow.

2.4. Data Analysis

Recorded values of ambient temperature, ventilation rate and ammonia concentration were integrated into daily average values presented for descriptive purposes.

The study data showed a normal distribution for the mean test estimates. The statistical analysis of each measured variable was conducted per week of age and accumulated throughout the growing period. A one-way ANOVA analysis was conducted using the software Satgraphics Centurion XVIII, according to the following model:

$$Yijk = \mu + Ti + \varepsilon i$$

where: Yijk = studied variable (week or accumulated value).

μ = general mean.

Ti = treatment (SHD Room 1, SHD Room 2, FHD Room 1, FHD Room 2, FLD Room 1, FLD Room 2).

εi = residual error.

3. Results and Discussion

3.1. Control of Environmental Parameters

Environment conditions (temperature, relative humidity, ammonia concentration and ventilation rate) during the experiment are presented in in the Table 1. Since both rooms had an identical number of animals and management, this difference in concentration was promoted through increasing ventilation in Room 1 as compared to Room 2.

Table 1. Weekly average values of temperature (°C), relative humidity (%) ventilation rate ($m^3\ h^{-1}$ animal^{-1}) and ammonia concentration (ppm) in the broiler houses (Room 1 and Room 2).

	Temperature (°C)		Relative Humidity (%)		Ventilation Rate ($m^3\ h^{-1}$ animal^{-1})		NH_3 Concentration (ppm)	
	Room 1	Room 2	Room 1	Room 1	Room 1	Room 2	Room 1	Room 2
Week 1	33.1	33.0	22.1	22.9	0.8	0.3	0.0	0.0
Week 2	29.8	29.5	29.4	31.7	4.2	3.7	0.0	0.0
Week 3	27.0	26.3	29.1	31.5	5.5	5.8	0.0	0.0
Week 4	24.7	23.8	29.5	30.5	6.5	7.7	0.0	0.0
Week 5	21.9	22.8	46.4	50.7	3.9	3.5	0.3	6.5
Week 6	20.3	22.0	54.1	59.2	3.5	2.5	4.5	19.0
Week 7	19.7	22.2	55.9	61.1	5.0	2.6	9.0	20.4
Week 8	20.4	21.7	48.4	50.8	6.5	3.7	12.5	14.4
Week 9	19.9	21.5	51.1	54.7	6.8	4.4	10.1	16.1

Air temperature (Table 1) followed the recommendations for the broiler strains used in this study. Both rooms had a similar temperature during the first four weeks. However, from week 5, the higher ventilation rates in Room 1 caused its average temperature to be reduced by 1.7 °C on average. However, the difference of 1.7 °C was obtained in the general average of all the weeks studied. In the first two weeks, which are the most critical, this difference did not reach 0.3 °C, thus being practically identical in temperature conditions is not enough to cause differences in the performance of the birds. Relative humidity was very similar between rooms. During the four initial weeks, ventilation was kept high, and near-zero ammonia concentration was detected (Table 1). From week 5 until the end of the experiment, the climate control system was operated to create the desired difference in ammonia concentration, averaging 8.27 ppm in Room 1 and 17.1 ppm in Room 2.

3.2. Animal Performance

The evolution of animal weight is presented in Table 2. As expected, slow-growing animals had lower weight than the fast-growing ones of the same age from the early stages of the growing period. During the first week, some differences were found among treatments, which could be related to the random choice of animals at the beginning of the experiment. Differences were also found in weeks 3 and 4 between slow-growing animals in both rooms. In those weeks, animals housed in Room 1 had a lower weight than those in Room 2. No more differences were found between slow-growing animals from day 35 until the end of the experiment. For fast-growing broilers, no differences in weight were detected between treatments from day 14 for animals kept at either high or low density.

Table 2. Evolution of animal weight (g per bird) during the rearing period (n = 30 animals per treatment).

Age	**Treatment SHD**		**Treatment FHD**		**Treatment FLD**		**S.E.**	***p*-Value**
(days)	Room 1	Room 2	Room 1	Room 2	Room 1	Room 2		
7	134.4 [a]	144.8 [ab]	164.4 [d]	151.1 [bc]	160.4 [cd]	160.5 [cd]	3.9	≤0.05
14	338.8 [a]	307.4 [a]	390.3 [b]	436.1 [b]	419.5 [b]	437.7 [b]	16.6	≤0.05
21	525.8 [a]	563.3 [b]	775.4 [c]	801.1 [c]	780.6 [c]	797.7 [c]	11.9	≤0.05
28	853.8 [a]	960.8 [b]	1242.7 [c]	1226.5 [c]	1239.2 [c]	1227.9 [c]	29.8	≤0.05
35	1294.1 [a]	1302.2 [a]	1840.8 [b]	1779.3 [b]	1825.9 [b]	1750.7 [b]	35.2	≤0.05
42	1769.0 [a]	1801.1 [a]	2527.3 [b]	2415.0 [b]	2475.6 [b]	2446.6 [b]	53.8	≤0.05
49	2192.0 [a]	2276.3 [a]	-	-	-	-	69.1	0.40
56	2676.8 [a]	2776.0 [a]	-	-	-	-	94.8	0.48
63	3127.7 [a]	3275.2 [a]	-	-	-	-	103.0	0.35

Different letters within a row indicate statistically significant differences among treatments. SHD = slow growing and high density; FHD = fast growing and high density; FLD = fast growing and low density. S.E. = standard error. Room 1 was programmed to reach maximum 10 ppm NH_3 and Room 2 to maximum 20 ppm NH_3.

Table 3 shows feed consumption per treatment and week. Among slow-growing animals, differences between rooms were only found from day 35 to day 42, when animals in Room 2 consumed more feed than in Room 1. No differences were found between rooms for fast-growing animals kept at a high density. However, feed consumption was higher in week 5 for low-density animals in Room 2 compared to Room 1. The accumulated feed consumption at the end of the growing period was higher for slow-growing animals (6.42 and 6.79 in Room 1 and Room 2, respectively) than for fast-growing animals, both at a high density (5.35 and 5.33 in Room 1 and Room 2, respectively) and at a low density (5.03 and 4.71 in Room 1 and Room 2, respectively). In summary, slow-growing animals consumed less feed per week but had higher feed consumption at the end of their growing period since their lifetime was three weeks longer (63 days for slow-growing animals and 42 days for fast-growing animals). Comparing fast-growing animals, those animals kept at a lower density and higher ammonia concentration (Room 2) had lower feed consumption at the end of the growing period.

Table 4 shows the evolution of feed conversion ratio depending on animal age. For the same treatment, no differences were found between Room 1 and Room 2. It was observed that both fast- and slow-growing animals kept at a high density had a similar feed conversion ratio at the end of their growing period. This means a proportional reduction in animal growth rate and feed consumption of slow-growing animals compared to fast-growing animals. However, fast-growing animals kept at a low density had a lower conversion ratio, particularly in those animals subjected to higher ammonia concentrations (Room 2). For the feed conversion ratio accumulated at 28 days of age, values of 1.90 were obtained for SHD, 2.00 for FHD, and 1.96 for FLD stocked in Room 1, and 1.66 for SHD, 2.01 for FHD and 1.82 for FLD stocked in Room 2. At 42 days of age, the following values were obtained: 1.76 for SHD, 2.16 for FHD and 2.09 for FLD stocked in Room 1, and 1.80 for SHD, 2.25 for FHD and 1.96 for FLD stocked in Room 2.

Table 3. Evolution of feed consumption (g per animal and week) during the rearing period (n = 30 animals per treatment).

Age (days)	Treatment SHD		Treatment FHD		Treatment FLD		S.E.	p-Value
	Room 1	Room 2	Room 1	Room 2	Room 1	Room 2		
7	0.116 ^{a}	0.113 ^{a}	0.136 ^{b}	0.132 ^{b}	0.132 ^{b}	0.128 ^{b}	0.004	≤0.05
14	0.256 ^{a}	0.255 ^{a}	0.368 ^{b}	0.373 ^{b}	0.350 ^{b}	0.370 ^{b}	0.008	≤0.05
21	0.518 ^{a}	0.490 ^{a}	0.732 ^{b}	0.693 ^{b}	0.703 ^{b}	0.627 ^{b}	0.037	≤0.05
28	0.672 ^{a}	0.664 ^{a}	1.155 ^{b}	1.184 ^{b}	1.135 ^{b}	1.038 ^{b}	0.068	≤0.05
35	0.677 ^{a}	0.682 ^{a}	1.452 ^{c}	1.445 ^{c}	1.430 ^{c}	1.114 ^{b}	0.048	≤0.05
42	0.814 ^{a}	0.978 ^{b}	1.506 ^{d}	1.506 ^{d}	1.287 ^{c}	1.438 ^{cd}	0.056	≤0.05
49	0.976 ^{a}	1.032 ^{a}	-	-	-	-	0.030	0.22
56	1.155 ^{a}	1.242 ^{a}	-	-	-	-	0.050	0.20
63	1.236 ^{a}	1.333 ^{a}	-	-	-	-	0.050	0.19

Different letters within a row indicate statistically significant differences among treatments. SHD = slow growing and high density; FHD = fast growing and high density; FLD = fast growing and low density. S.E. = standard error. Room 1 was programmed to reach maximum 10 ppm NH_3 and Room 2 to maximum 20 ppm NH_3.

Table 4. Evolution of feed conversion ratio per week during the rearing period (n = 30 animals per treatment).

Age (days)	Treatment SHD		Treatment FHD		Treatment FLD		S.E.	p-Value
	Room 1	Room 2	Room 1	Room 2	Room 1	Room 2		
7	1.21 ^{ab}	1.07 ^{a}	1.12 ^{ab}	1.27 ^{b}	1.13 ^{ab}	1.10 ^{a}	0.05	0.10
14	1.56 ^{b}	1.59 ^{b}	1.38 ^{a}	1.32 ^{a}	1.36 ^{a}	1.39 ^{a}	0.05	≤0.05
21	2.22 ^{b}	1.94 ^{ab}	2.11 ^{ab}	1.92 ^{ab}	2.01 ^{ab}	1.77 ^{a}	0.12	0.17
28	2.01 ^{ab}	1.74 ^{a}	2.53 ^{c}	2.79 ^{c}	2.53 ^{bc}	2.39 ^{bc}	0.18	≤0.05
35	1.60 ^{a}	2.06 ^{ab}	2.44b ^{c}	2.63 ^{c}	2.45 ^{bc}	2.09 ^{ab}	0.18	≤0.05
42	1.72 ^{a}	2.06 ^{ab}	2.20 ^{b}	2.30 ^{b}	2.02 ^{ab}	2.16 ^{b}	0.12	≤0.05
49	2.46 ^{a}	2.28 ^{a}	-	-			0.26	0.63
56	2.39 ^{a}	2.51 ^{a}	-				0.07	0.25
63	2.76 ^{a}	2.69 ^{a}	-				0.11	0.66

Different letters within a row indicate statistically significant differences among treatments. SHD = slow growing and high density; FHD = fast growing and high density; FLD = fast growing and low density. Room 1 was programmed to reach maximum 10 ppm NH^3 and Room 2 to maximum 20 ppm NH_3.

A different exposure of ammonia between Room 1 and Room 2 was achieved using different ventilation management strategies. Although the experiment was planned in winter conditions, where lower ventilation rates are needed, ammonia concentration was negligible until day 32 of the growing period. Litter material remained too dry to generate significant NH_3 emissions, so on day 32, urea had to be applied in both rooms to increase ammonia production. As the same rate of urea was used in both rooms, it was not expected to disturb the results of the experiment and was effective in producing a controlled amount of ammonia in the rooms. Once ammonia was generated, the differential effect of ventilation originated different ammonia concentrations until the end of the growing period. Therefore, exposure to differential ammonia occurred during the last 10 days for the fast-growing animals and the second half of the growing period for the slow-growing animals. These conditions may be representative of broiler production if new bedding material is used, since concentrations are low during the first weeks of the rearing period [19].

Ventilation operated by the ammonia sensor was effective to control ammonia concentrations. Until day 32, both rooms operated with similar ventilation (less than 2% difference on average). However, from that day, Room 1 had on average 59% higher ventilation than Room 2. This is concurrent with the lower ammonia concentration registered, which was 51% lower in Room 1.

Ventilation has been reported as an effective way to reduce indoor ammonia concentration; this strategy involves more energy consumption. Was r eported that a commercial farm could keep gas concentrations below regulation thresholds (20 ppm) by increasing ventilation, but this would involve increasing electric and heat consumption by 10% and 14%, respectively. Using ventilation to keep lower concentration values would entail higher energy costs as well as instrument maintenance, and these costs would be likely increase exponentially as the target value reduces. Therefore, efforts to keep concentrations lower than those reported in the regulation must be clearly supported in terms of productivity, health and welfare [18].

In this study, we found no significant effect of exposure to ammonia concentration on broilers kept under conventional conditions (fast-growing animals kept at high stocking density). Animals under improved environmental conditions tended to grow faster with similar feed consumption, which resulted in a slightly better conversion rate. These numerical differences are in accordance with the hypothesis of this study. Depending on the stress factor, the organs related to immunity would decrease or increase in size. In the case of stress due to ammonia exposure, the liver, spleen and bursa would be less developed [20]. However, no differences were found in this study.

Exposure to differential ammonia concentrations for only 10 days could be originating that differences are minor and could not be detected in this experiment. The effects of ammonia on animals depend on exposure time [4], and therefore it is necessary to explore the effect of longer exposure times, which could also affect broilers at early stages of their life.

An ammonia concentration below 20 ppm does not appear to have a clear effect on productive parameters, and similar results have been found in other experiments [11,21]. It seems that exposure to low ammonia concentrations may influence some conditions related to health and welfare [12], but this disturbance is not enough to produce visible changes in performance. According to our results, it seems that the absence of an effect of low ammonia concentrations on productive parameters would also apply for slow-growing strains and animals kept at a lower stocking density.

This study also did not find a significant effect of stocking density on animal growth in fast-growing animals. However, we found that animals reduced feed consumption and improved the feed conversion ratio, particularly for animals kept under higher ammonia concentrations. Improvements of lowering stocking density have been reported in the literature [22,23], but the interacting effect of ammonia concentration has not been reported. It seems that reducing stocking density may facilitate animals responding better to adverse environmental conditions, although the high-density treatment in this study was not sufficiently high to elicit more relevant differences at the animal level, as those obtained in previous studies [15,24].

The difference in productive parameters between slow-growing and fast-growing strains was evident. However, feed conversion rate did not decrease when using the slow-growing strain used in this experiment, which is relevant in terms of efficiency in the use of resources. Slow-growing animals were exposed for more time (31 days) to different ammonia concentrations than fast-growing animals (only 10 days), and therefore more differences would be expected. However, very similar productive parameters were found for slow-growing animals kept either at 10 or 20 ppm. The only difference detected was the relative weight of spleen at the end of the growing period. The lower spleen weight in animals kept at higher ammonia concentrations would support the hypothesis that stressor factors reduce immune organs [20], but this would not be translated into different productive results.

The results of this study indicate that husbandry practices can improve the conditions of animal growth with respect to the minimum legal requirements. These improved conditions, however, have a relatively low impact on animal performance and therefore their cost effectiveness needs to be further assessed. Results suggest that exposure to low ammonia concentrations (10 vs. 20 ppm) during part of the growing period originates some physiological changes that are not clearly reflected in productive terms, and we did not find evidence that these improved conditions could allow animals to respond better to stressing conditions. This potential effect on animal resiliency needs to be explored by future research.

3.3. Weight of the Immune Organs

The effect of stress on the welfare of the animal, its health and therefore its immune system is varied. The spleen, liver, and bursa of Fabricius are used as anatomical indicators of stress [25,26]. The average weight of the lymphoid organs (spleen and bursa of Fabricius) is related to changes in response to stress. The same has been reported for the thymus [27]. In Table 5, we find the evolution of the ratio of the organs that are involved in the immune system and are indicators of stress. Depending on the stress factor, the organ decreases or increases its size. In the case of stress due to a high concentration of NH_3, the organs are expected to be underdeveloped compared to a normal growth [20]. Contrarily, the expected effect of density is that the immune organs are more developed than normal [15].

Table 5. Evolution of the relative weight of spleen, thymus, liver and bursa Fabricius (g per 1000 g animal weight) during the rearing period.

Organ	Age	Treatment SHD		Treatment FHD		Treatment FLD		*p*-Value
	(days)	Room 1	Room 2	Room 1	Room 2	Room 1	Room 2	
Spleen	21	1.11 ± 1.17	1.27 ± 1.24	3.82 ± 1.19	0.94 ± 1.17	0.91 ± 1.19	0.94 ± 1.17	0.47
	42	1.62 ^{b} ± 0.06	1.46 ^{b} ± 0.06	1.25 ^{a} ± 0.06	1.18 ^{a} ± 0.06	1.46 ^{b} ± 0.06	1.21 ^{a} ± 0.06	≤0.05
	63	1.26 ^{b} ± 0.05	1.10 ^{a} ± 0.05	-	-	-	-	≤0.05
Thymus	21	2.88 ^{c} ± 0.15	2.98 ^{c} ± 0.16	1.85 ^{ab} ± 0.16	2.04 ^{b} ± 0.15	1.75 ^{ab} ± 0.16	1.57 ^{a} ± 0.15	≤0.05
	42	2.90 ^{b} ± 0.16	2.71 ^{b} ± 0.16	1.59 ^{a} ± 0.16	1.80 ^{a} ± 0.16	1.81 ^{a} ± 0.16	1.79 ^{a} ± 0.16	≤0.05
	63	2.44 ± 0.16	2.22 ± 0.16	-	-	-	-	0.32
Liver	21	29.89 ^{c} ± 0.79	29.38 ^{bc} ± 0.83	26.69 ^{a} ± 0.80	27.71 ^{abc} ± 0.79	27.50 ^{ab} ± 0.80	28.35 ^{abc} ± 0.79	≤0.05
	42	20.55 ^{a} ± 0.48	21.69 ^{a} ± 0.48	20.83 ^{a} ± 0.48	21.56 ^{a} ± 0.48	21.76 ^{a} ± 0.48	21.25 ^{a} ± 0.48	0.38
	63	17.68 ± 0.52	18.69 ± 0.52	-	-	-	-	0.18
Bursa of Fabricius	21	2.85 ^{b} ± 0.13	2.33 ^{a} ± 0.14	1.98 ^{a} ± 0.13	2.09 ^{a} ± 0.13	2.12 ^{a} ± 0.13	2.33 ^{a} ± 0.13	≤0.05
	42	0.54 ^{bc} ± 0.05	0.44 ^{abc} ± 0.05	0.42 ^{ab} ± 0.05	0.45 ^{abc} ± 0.05	0.40 ^{a} ± 0.05	0.56 ^{c} ± 0.05	0.08
	63	-	-	-	-	-	-	-

Different letters within a row indicate statistically significant differences among treatments. SHD = slow growing and high density; FHD = fast growing and high density; FLD = fast growing and low density. Room 1 was programmed to reach maximum 10 ppm NH_3 and Room 2 to maximum 20 ppm NH_3.

Table 5 shows the weight of the organs related to the immune system, expressed in relative terms with respect to total animal weight. For fast-growing animals, information is available until the slaughter day (42). For slow-growing animals, the bursa of Fabricius was too small, and it could not be measured at day 63, so no information is provided.

Spleen weight was very variable on day 21 of the growing period and no relevant differences were found on day 42 in absolute weight. However, as shown in Table 5, the spleen had a higher weight when expressed in relative terms to animal weight due to the lower weight of slow-growing animals at the same age. Additionally, animals at a higher ammonia concentration tended to have a lower spleen weight than those at lower concentrations. At slaughter age, slow-growing animals' spleens were heavier in animals kept at lower ammonia concentration (Room 1) compared to those at higher ammonia concentrations (Room 2). The relative weight of the thymus gland was also affected by animal strain and slow-growing animals had a higher relative weight than fast-growing animals. No differences for stocking density were detected due to ammonia concentration. A similar situation was found for the liver and bursa of Fabricius, where no clear effect was found because of changing exposure to ammonia concentration or stocking density.

Studies show that in situations of stress due to exposure to ammonia, the liver, spleen and bursa would be less developed [20,28,29]. This study found some changes in the size of immune organs, so it is suggested that more in-depth studies are necessary in this direction to confirm if in fact the larger size of the organs are correlated with situations of greater stress such as higher ammonia concentration and higher stocking density.

4. Conclusions

This study evaluated the effect of husbandry practices on broiler response to low ammonia concentrations. Ventilation was used to generate exposure to different ammonia concentration in the final part of a growing period (approximately 10 and 20 ppm).

Reducing ammonia concentrations had a low influence on productive parameters regardless of the animal strain (slow- or fast-growing animals) or stocking density (32 vs. 16 kg m^{-2} at slaughter age) was tested. However, animals kept at a lower stocking density had a better conversion rate than those under high density, particularly at higher ammonia exposure.

Slow-growing animals were exposed to an ammonia concentration for a longer time than fast-growing animals. No effects on growth parameters were found, but the increased weight of immune organs (spleen) suggests a deleterious effect on the immune response. However, despite being an indication, it is not possible to say that there is an immunological dysfunction based only on the weight of the lymphoid organs. Therefore, we recommend that further research be carried out in this direction.

Husbandry practices improving animal conditions beyond the minimum legal requirements seem to have a low effect on productivity.

Author Contributions: Conceptualization, L.V.S.B., S.C., A.R.-M., F.E. and A.V.; methodology, S.C., F.E. and A.V.; software, L.V.S.B.; A.R.-M. and A.V.; validation, L.V.S.B., S.C., F.E. and A.R.-M.; formal analysis, L.V.S.B., D.J.D.M., F.E., A.R.-M., S.C. and A.V.; investigation, L.V.S.B., D.J.D.M., S.C., A.R.-M., F.E. and A.V.; data curation, L.V.S.B., S.C., A.R.-M. and F.E.; writing—original draft preparation, L.V.S.B., D.J.D.M. and S.C.; writing—review and editing, L.V.S.B., D.J.D.M., F.E., S.C. and A.V.; supervision, L.V.S.B., A.R.-M. and A.V.; project administration, A.V., F.E. and S.C. and took the photos, L.V.S.B. and A.R.-M. All figures were drawn by the authors. All authors have read and agreed to the published version of the manuscript.

Funding: This research received no external funding.

Institutional Review Board Statement: The experiment was conducted in accordance with the animal research regulations of the EU, with protocol number 2018/VSC/PEA/0067. The test was carried out at the Animal Technology and Research Center (CITA-IVIA), located in Segorbe, (Castellón, Spain).

Informed Consent Statement: Not applicable.

Data Availability Statement: Not applicable.

Acknowledgments: This research was funded by the National Institute for Agricultural Research and Experimentation and the Ministry of Economy, Industry and Competitiveness (RTA 2017-00013, Programme: MINECO, Ministerio de Economía y Competitividad). The scholarship Capes and Cnpq supported the participation of Leonardo Valentino in this research. We acknowledge CNPq for the participation number 308177/2021-5 and Santander for the Subject Ibero-Americana.

Conflicts of Interest: The authors declare no conflict of interest.

References

1. Tasistro, A.S.; Ritz, C.W.; Kissel, D.E. Ammonia emissions from broiler litter: Response to bedding materials and acidifiers. *Br. Poult. Sci.* **2007**, *48*, 399–405. [CrossRef] [PubMed]
2. Kearney, G.D.; Shaw, R.; Prentice, M.; Tutor-Marcom, R. Evaluation of Respiratory Symptoms and Respiratory Protection Behavior Among Poultry Workers in Small Farming Operations. *J. Agromed.* **2014**, *19*, 162–170. [CrossRef] [PubMed]
3. Nemer, M.; Sikkeland, L.I.B.; Kasem, M.; Kristensen, P.; Nijem, K.; Bjertness, E.; Skare, Ø.; Bakke, B.; Kongerud, J.; Skogstad, M. Airway inflammation and ammonia exposure among female Palestinian hairdressers: A cross-sectional study. *Occup. Environ. Med.* **2015**, *72*, 428–434. [CrossRef]
4. Kristensen, H.H.; Wathes, C.M. Ammonia and poultry welfare: A review. *Worlds. Poult. Sci. J.* **2000**, *56*, 235–245. [CrossRef]
5. Beker, A.; Vanhooser, S.L.; Swartzlander, J.H.; Teeter, R.G. Atmospheric ammonia concentration effects on broiler growth and performance. *J. Appl. Poult. Res.* **2004**, *13*, 5–9. [CrossRef]
6. Miles, D.M.; Branton, S.L.; Lott, B.D. Atmospheric ammonia is detrimental to the performance of modern commercial broilers. *Poult. Sci.* **2004**, *83*, 1650–1654. [CrossRef]

7. Miles, D.M.; Miller, W.W.; Branton, S.L.; Maslin, W.R.; Lott, B.D. Ocular Responses to Ammonia in Broiler Chickens. *Avian Dis.* **2006**, *50*, 45–49. [CrossRef]
8. Purswell, J.L.; Davis, J.D.; Kiess, A.S.; Coufal, C.D. Effects of frequency of multiple applications of litter amendment on litter ammonia and live performance in a shared airspace. *J. Appl. Poult. Res.* **2013**, *22*, 469–473. [CrossRef]
9. Wei, F.X.; Hu, X.F.; Xu, B.; Zhang, M.H.; Li, S.Y.; Sun, Q.Y.; Lin, P. Ammonia concentration and relative humidity in poultry houses affect the immune response of broilers. *Genet. Mol. Res.* **2015**, *14*, 3160–3169. [CrossRef]
10. Yi, B.; Chen, L.; Sa, R.; Zhong, R.; Xing, H.; Zhang, H. Transcriptome profile analysis of breast muscle tissues from high or low levels of atmospheric ammonia exposed broilers (Gallus gallus). *PLoS ONE* **2016**, *11*, e0162631. [CrossRef]
11. Zhou, Y.; Liu, Q.X.; Li, X.M.; Ma, D.D.; Xing, S.; Feng, J.H.; Zhang, M.H. Effects of ammonia exposure on growth performance and cytokines in the serum, trachea, and ileum of broilers. *Poult. Sci.* **2020**, *99*, 2485–2493. [CrossRef] [PubMed]
12. Zhou, Y.; Zhang, M.; Liu, Q.; Feng, J. The alterations of tracheal microbiota and inflammation caused by different levels of ammonia exposure in broiler chickens. *Poult. Sci.* **2021**, *100*, 685–696. [CrossRef] [PubMed]
13. Rayner, A.C.; Newberry, R.C.; Vas, J.; Mullan, S. Slow-growing broilers are healthier and express more behavioural indicators of positive welfare. *Sci. Rep.* **2020**, *10*, 1–14. [CrossRef] [PubMed]
14. Manswr, B.; Ball, C.; Forrester, A.; Chantrey, J.; Ganapathy, K. Host immune response to infectious bronchitis virus Q1 in two commercial broiler chicken lines. *Res. Vet. Sci.* **2021**, *136*, 587–594. [CrossRef]
15. Abouelenien, F.; KhalfAlla, F.; MousaBalabel, T.; Nasser, S. Effect of Stocking Density and Bird Age on Air Ammonia, Performance and Blood Parameters of Broilers. *Worlds Vet. J.* **2016**, *6*, 130. [CrossRef]
16. Ferguson, N.S.; Gates, R.S.; Taraba, J.L.; Cantor, A.H.; Pescatore, A.J.; Ford, M.J.; Burnham, D.J. The Effect of Dietary Crude Protein on Growth, Ammonia Concentration, and Litter Composition in Broilers. *Poult. Sci.* **1998**, *77*, 1481–1487. [CrossRef]
17. Miles, D.M.; Rowe, D.E.; Cathcart, T.C. Litter ammonia generation: Moisture content and organic versus inorganic bedding materials. *Poult. Sci.* **2011**, *90*, 1162–1169. [CrossRef]
18. Costantino, A.; Fabrizio, E.; Villagrá, A.; Estellés, F.; Calvet, S. The reduction of gas concentrations in broiler houses through ventilation: Assessment of the thermal and electrical energy consumption. *Biosyst. Eng.* **2020**, *199*, 135–148. [CrossRef]
19. Calvet, S.; Cambra-López, M.; Estellés, F.; Torres, A.G. Characterization of gas emissions from a Mediterranean broiler farm. *Poult. Sci.* **2011**, *90*, 534–542. [CrossRef]
20. Wang, Y.M.; Meng, Q.P.; Guo, Y.M.; Wang, Y.Z.; Wang, Z.; Yao, Z.L.; Shan, T.Z. Effect of Atmospheric Ammonia on Growth Performance and Immunological Response of Broiler Chickens. *J. Anim. Vet. Adv.* **2010**, *22*, 2802–2806. [CrossRef]
21. Zarnab, S.; Chaudhary, M.S.; Javed, M.T.; Khatoon, A.; Saleemi, M.K.; Ahmed, T.; Tariq, N.; Manzoor, F.; Javed, I.; Zhang, H.; et al. Effects of induced high ammonia concentration in air on gross and histopathology of different body organs in experimental broiler birds and its amelioration by different modifiers. *Pak. Vet. J.* **2019**, *39*, 371–376. [CrossRef]
22. Li, X.M.; Zhang, M.H.; Liu, S.M.; Feng, J.H.; Ma, D.D.; Liu, Q.X.; Zhou, Y.; Wang, X.J.; Xing, S. Effects of stocking density on growth performance, growth regulatory factors, and endocrine hormones in broilers under appropriate environments. *Poult. Sci.* **2019**, *98*, 6611–6617. [CrossRef] [PubMed]
23. Liu, H.; Bai, S.P.; Zhang, K.Y.; Ding, X.M.; Wang, J.P.; Zeng, Q.F.; Peng, H.W.; Bai, J.; Xuan, Y.; Su, Z.W. Effects of stocking density on the performance, tibia mineralization, and the expression of hypothalamic appetite genes in broiler chickens. *Ann. Anim. Sci.* **2021**, *21*, 1103–1117. [CrossRef]
24. Buijs, S.; Keeling, L.; Rettenbacher, S.; van Poucke, E.; Tuyttens, F.A.M. Stocking density effects on broiler welfare: Identifying sensitive ranges for different indicators. *Poult. Sci.* **2009**, *88*, 1536–1543. [CrossRef]
25. Pope, C.R. Pathology of lymphoid organs with emphasis on immunosuppression. *Vet. Immunol. Immunopathol.* **1991**, *30*, 31–44. [CrossRef]
26. Freire, R.; Wilkins, L.J.; Short, F.; Nicol, C.J. Behaviour and welfare of individual laying hens in a non-cage system. *Br. Poult. Sci.* **2003**, *44*, 22–29. [CrossRef]
27. Ravindran, V.; Thomas, D.V.; Thomas, D.G.; Morel, P.C.H. Performance and welfare of broilers as affected by stocking density and zinc bacitracin supplementation. *Anim. Sci. J.* **2006**, *77*, 110–116. [CrossRef]
28. Xing, H.; Luan, S.; Sun, Y.; Sa, R.; Zhang, H. Effects of ammonia exposure on carcass traits and fatty acid composition of broiler meat. *Anim. Nutr.* **2016**, *2*, 282–287. [CrossRef]
29. Soliman, E.S.; Hassan, R.A. Evaluation of superphosphate and meta-bisulfide efficiency in litter treatment on productive performance and immunity of broilers exposed to ammonia stress. *Adv. Anim. Vet. Sci.* **2017**, *5*, 253–259. [CrossRef]

animals

MDPI

Article

Biomarkers and *De Novo* Protein Design Can Improve Precise Amino Acid Nutrition in Broilers

María Cambra-López [1,*], Pablo Jesús Marín-García [2], Clara Lledó [1], Alba Cerisuelo [3] and Juan José Pascual [1]

1 Institute for Animal Science and Technology, Universitat Politècnica de València, Camino de Vera s/n, 46022 Valencia, Spain; claralledomorell@outlook.es (C.L.); jupascu@dca.upv.es (J.J.P.)
2 Departamento Producción y Sanidad Animal, Salud Pública y Ciencia y Tecnología de los Alimentos, Facultad de Veterinaria, Universidad Cardenal Herrera-CEU, CEU Universities, 46113 Valencia, Spain; pablo.maringarcia@uchceu.es
3 Centro de Investigación de Tecnología Animal, Instituto Valenciano de Investigaciones Agrarias, 12400 Segorbe, Spain; cerisuelo_alb@gva.es
* Correspondence: macamlo@upvnet.upv.es

Simple Summary: Almost half of the protein ingested by broilers is not retained and is excreted, impairing the nitrogen utilization, health and productivity of the animals, and intensifying the environmental impact of poultry meat production. This work proposes two potential tools, combining traditional nutrition with biotechnological, metabolomics, computational and protein engineering knowledge, which can contribute to improving precise amino acid nutrition in broilers in the future: (i) the use of serum uric nitrogen content as a rapid biomarker of amino acid imbalances, and (ii) the design and modeling of *de novo* proteins that are fully digestible and fit exactly to the animal's requirements. Both tools can open up new opportunities to form an integrated framework for precise amino acid nutrition in broilers, helping us to achieve more efficient, resilient, and sustainable production. This information can help to determine the exact ratio of amino acids that will improve the efficiency of the use of nitrogen by poultry.

Abstract: Precision nutrition in broilers requires tools capable of identifying amino acid imbalances individually or in groups, as well as knowledge on how more digestible proteins can be designed for innovative feeding programs adjusted to animals' dynamic requirements. This work proposes two potential tools, combining traditional nutrition with biotechnological, metabolomic, computational and protein engineering knowledge, which can contribute to improving the precise amino acid nutrition of broilers in the future: (i) the use of serum uric nitrogen content as a rapid biomarker of amino acid imbalances, and (ii) the design and modeling of *de novo* proteins that are fully digestible and fit exactly to the animal's requirements. Each application is illustrated with a case study. Case study 1 demonstrates that serum uric nitrogen can be a useful rapid indicator of individual or group amino acid deficiencies or imbalances when reducing dietary protein and adjusting the valine and arginine to lysine ratios in broilers. Case study 2 describes a stepwise approach to design an ideal protein, resulting in a potential amino acid sequence and structure prototype that is ideally adjusted to the requirements of the targeted animal, and is theoretically completely digestible. Both tools can open up new opportunities to form an integrated framework for precise amino acid nutrition in broilers, helping us to achieve more efficient, resilient, and sustainable production. This information can help to determine the exact ratio of amino acids that will improve the efficiency of the use of nitrogen by poultry.

Keywords: precision livestock farming; PLF; precise feeding; poultry; ideal protein

Citation: Cambra-López, M.; Marín-García, P.J.; Lledó, C.; Cerisuelo, A.; Pascual, J.J. Biomarkers and *De Novo* Protein Design Can Improve Precise Amino Acid Nutrition in Broilers. *Animals* **2022**, *12*, 935. https://doi.org/10.3390/ani12070935

Academic Editor: Alireza Seidavi

Received: 28 February 2022
Accepted: 29 March 2022
Published: 6 April 2022

1. Introduction

Precision nutrition is not a new concept. It was used for the first time in poultry nutrition in 1979 in a precision feeding bioassay to measure true available amino acids in roosters [1]. Precision nutrition combines traditional nutrition with other disciplines (mathematics, computer sciences, chemistry, biochemistry, biology, immunology, molecular biology, genetics, engineering and technological sciences, amongst others) in a multidisciplinary approach [2].

It can be defined as the practice of meeting the nutrient requirements of animals as accurately as possible in the interests of safe, high-quality and efficient production, while ensuring the lowest possible load on the environment [3]. Therefore, it aims at precisely matching animals' nutritional requirements with adjusted feed diets, and requires a well-characterized and accurate nutrient database for each ingredient, together with properly defined animal nutrient requirements [4].

By definition, this concept is inherently linked to animal farming practices, and is key to optimizing feed efficiency for maximal economic return and minimum losses. However, despite its history of use, paradoxically, the practical implementation of precise nutrition in broiler production is not yet entirely achieved. In fast-growing broilers, nutritional requirements change quickly over time, and daily variations cannot be met with multiphase-feeding only [5,6], or by blending diets [4,7].

Nutritional requirements are commonly set for a population of similar animals (according to their age, physiological status and/or genetics, and occasionally sex). Using the population-feeding approach, individual variations within animals cannot be addressed [8], and singular needs according to nutritional status, genetics or animal health, and environmental stress-related conditions may be consequently overlooked. Managing animals individually is key in precision livestock farming [9], and can prevent over-feeding, particularly in growing pigs [8]. Nevertheless, it is questionable whether the precision management of individual birds is feasible in the poultry sector [2].

Protein over-feeding results in an increasing nitrogen (N) environmental load and ammonia emissions, and causes economic losses [10]. Birds need adjusted amino acid levels that are ideally combined (using the ideal protein concept, expressed relative to lysine [11]) to meet the requirements of each amino acid without deficit or excess. Even though broilers are one of the most efficient animals in transforming proteins into meat, compared with swine or cattle [12], their N retention is low and ranges from 57 to 60% [13]. Therefore, almost half of the protein ingested by broilers is not retained and is excreted.

Moreover, undigested protein and the metabolites from protein fermentation (ammonia, amines, p-cresol and indole [14]) can negatively affect intestinal health [15]. Undigested protein in the distal gastrointestinal tract can disrupt gut function and integrity [16], and can also be used by undesirable pathogenic bacteria [17]. Furthermore, if amino acids are available in excess or improperly balanced, they need to be catabolized in the liver. As a consequence, ammonia, which is highly toxic, is produced and must be released. The deamination of unused amino acids in the liver and the excretion of ammonia as uric acid in poultry is costly for the animal, requiring a supply of energy in the form of adenosine triphosphate (ATP)—three ATP molecules are consumed for every N molecule excreted [18]. All this seriously worsens the health and productivity of the animals and intensifies the environmental impact of poultry meat production.

Even though nowadays nutritionists use optimized feed supply and animal amino acid requirement evaluation methods (based on true ileal digestible amino acids, rather than traditional crude protein estimation or fecal total amino acids; together with modeling approaches that can assist in the process [19,20]), further research on balanced feeds, with maximal amino acid digestibility tailored for each animal's requirements over time, is needed.

To achieve the ideal fitting of amino acid supply to animals' dynamic requirements, precision nutrition in broilers requires tools capable of identifying deficiencies or imbalances individually or in groups, as well as knowledge on how more digestible and usable proteins can be obtained or even designed for innovative feeding programs. This new framework would help reduce the detrimental effects of protein over-feeding and inaccurate amino acid balancing in broiler diets.

This work proposes two potential tools, combining traditional nutrition with biotechnological, metabolomic, computational and protein engineering knowledge, which can contribute to improving precise amino acid nutrition in broilers in the future: (i) the use of serum uric N (SUN) content as a rapid biomarker of amino acid imbalances, and (ii) the design and modeling of *de novo* proteins that are fully digestible and fit exactly to the animal's requirements. Each application is illustrated with a case study. The required future improvements in protein nutrition using precision nutrition tools in broilers are further discussed.

2. Case Study 1: Use of SUN Content as a Rapid Biomarker of Amino Acid Imbalances

2.1. Background

Serum uric nitrogen corresponds to the amount of N in the form of uric acid circulating in the bird's bloodstream. Therefore, it can be used as a metabolic indicator of amino acid imbalances and deficiencies. This biomarker has become more common in the last few decades as a valid criterion to determine amino acid requirements under conditions of constant protein intake [21]. It has been successfully validated in swine [21–23], rabbits [24] and broilers [25].

In sows fed a diet that is adequate in the first-limiting amino acids, the concentration of plasma urea nitrogen is low because there is a decrease in protein catabolism, more efficient total N utilization, and thus a decrease in urea synthesis [21]. In rabbits, a diet with an imbalance in any essential amino acid would lead animals to catabolize the remaining amino acids, increasing the urea production in the liver, which would be released into the bloodstream, increasing plasma urea nitrogen [24]. The higher the excess of digested protein and the more limiting the affected amino acid is, the higher the plasma or SUN levels will be.

Methionine followed by lysine are the first limiting amino acids in most practical poultry diets [26–28], and their requirements are generally accurately estimated [29–32]. The requirements of other amino acids, however, still need further adjustment. For example, the recommendations on the valine and arginine to lysine ratios are near 0.80 [33] and 1.05 [34], respectively. Their high nutritional requirements and their relative low presence in commercial diets indicate they are relevant amino acids, which may become limiting in specific situations. Therefore, there is a need to determine their levels accurately in broiler diets to optimize both growth and N use.

This case study illustrates how SUN content can be used as a valid biomarker to detect imbalances and deficiencies in secondary limiting amino acids in broilers. The use of this biomarker is a promising tool used to verify feed formulations, monitor the ideal balancing of amino acids in broilers, and aid in adjusting amino acids to precisely match animal's requirements over time.

A trial was conducted to determine the effects of reducing dietary protein and adjusting valine and arginine to lysine ratios in broilers. The level of SUN metabolite was used to identify potential amino acid imbalances. The relationship between SUN and performance traits was also evaluated.

2.2. Animals and Experimental Procedure

Three hundred and thirty-six male broilers (Ross 308) were assigned to four dietary treatments from days 14 to 35 of age. Before that, all birds were fed a commercial diet. Animals were reared in floor pens in an environmentally controlled room.

All experimental procedures used in this study were approved by Universitat Politècnica de València's Animal Experimentation Ethics Committee, and authorized by the Valencian Conselleria de Agricultura, Medio Ambiente, Cambio Climático y Desarrollo of Spain, with the code 2017/VSC/PEA/000166.

There were seven pens per treatment (1.3 m × 1.3 m) and 12 animals/pen. Diets were formulated to meet the birds' crude protein requirements (20%; in T1) or to be below the crude protein requirements (18%; in T2, T3 and T4) [35], combined with changes in valine (0.70 to 0.80) and arginine (0.90 to 1.05) to lysine ratios.

Diets were formulated to be isoenergetic (3000 kcal metabolizable energy/kg) and pelleted (target pellet temperature = 70 °C). The valine to lysine ratio was formulated according to current recommendations (average analyzed value of 0.81; Table 1) in T1, T2 and T4, and it was below these recommendations in T3 (0.71; Table 1). The arginine to lysine ratio was formulated according to current recommendations (average analyzed value of 1.07; Table 1) in T1, T2 and T3, and was below them in T4 (0.93; Table 1). Amino acid changes in dietary treatments were established by adding synthetic amino acids to a common basal diet based on corn, wheat and soybean meal.

Table 1. Analyzed levels of crude protein, valine and arginine, valine to lysine ratio and arginine to lysine ratio in the different dietary treatments (T1 to T4).

	Crude Protein (%)	Valine (%)	Arginine (%)	Valine to Lysine	Arginine to Lysine
T1	20.00	0.896	1.228	0.815	1.116
T2	18.13	0.873	1.136	0.806	1.049
T3	18.07	0.770	1.136	0.711	1.049
T4	17.88	0.874	1.004	0.807	0.927

Different levels of valine and arginine were achieved by adding synthetic L-valine and L-arginine. T1: 20% crude protein content, valine/lysine ratio (0.80) and arginine/lysine ratio (1.05) formulated according to current recommendations; T2: 18% crude protein content and valine/lysine ratio (0.80) and arginine/lysine ratio (1.05) formulated according to current recommendations; T3: 18% crude protein content, below-required valine/lysine ratio (0.70) and arginine/lysine ratio (1.05) formulated according to current recommendations; T4: 18% crude protein content, below-required arginine/lysine ratio (0.92) and valine/lysine ratio (0.80) formulated according to current recommendations.

Individual body weight and pen feed intake were controlled on days 14, 21, 28 and 35 of age. On day 36 of age, animals were fasted for 2 h, and blood samples were obtained 90 min after giving them access to feed. Blood samples were obtained from the wing veins in 84 animals (2115 ± 11 g body weight; 3 animals per pen; 21 animals per treatment) in 3 mL serum tubes (vacutainers). Blood samples were centrifuged (10 min, 3000× g) and stored frozen (−20 °C) until analyses.

The determination of SUN was performed using a commercial kit (Urea/BUN-Color, BioSystems S.A., Barcelona, Spain). Samples were firstly defrosted and tempered, and then 1 µL was pipetted into test tubes (a standard and a blank were included in each batch). Later, 1 mL of reagent A (sodium salicylate 62 mmol/L, sodium nitroprusside 3.4 mmol/L, phosphate buffer 20 mmol/L and urease 500 U/mL) was added to each sample, mixed thoroughly and incubated for 5 min at 37 °C. Subsequently, 1 mL of reactant B (sodium hypochlorite 7 mmol/L and sodium hydroxide 150 mmol/L) was added, mixed thoroughly and incubated for a further 5 min at 37 °C. Finally, the absorbance of each sample was read at 600 nm against the blank.

Individual bird SUN, final bird body weight and average daily gain (ADG) data were statistically analyzed with the GLM procedure of the SAS System Software®. The experimental diet (T1 to T4) was considered as the fixed effect in the model. Least square means were obtained with standard errors. Significant differences were declared at $p \leq 0.05$.

2.3. Results

Table 2 shows the SUN values and productive traits (mean ± standard error of the mean) obtained for each experimental diet. The average SUN varied from 1.89 ± 0.1 to 2.26 ± 0.1 mg/dL in animals fed the tested diets. Animals fed diet T4 showed the highest SUN values (on average +18%, $p < 0.05$) compared with groups T1 to T3. These results agree with the performance data (weight and ADG), where T4 showed lower values compared with T1 and T2 ($p < 0.05$). On the other hand, the SUN concentrations were similar among treatments T1 to T3. The final weight and ADG were the highest in animals fed diet T1, medium in diet T2, and the lowest in animals fed diets T3 and T4.

Table 2. Average serum uric nitrogen (SUN) and productive traits (± standard error of the mean) obtained in each experimental diet (T1 to T4) during the grower phase (from 14 to 35 days of age).

	T1	T2	T3	T4	*p*-Value
SUN (mg/dL) [1]	1.96 ± 0.1 [a]	1.89 ± 0.1 [a]	1.90 ± 0.1 [a]	2.26 ± 0.1 [b]	0.001
Final weight (g)	2188 ± 18.4 [c]	2143 ± 18.4 [b]	2073 ± 18.4 [a]	2055 ± 18.4 [a]	0.028
Average daily gain (g/d)	109.3 ± 1.62 [c]	105.7 ± 1.62 [b]	100.8 ± 1.62 [a]	101.7 ± 1.62 [a]	0.043

[a, b, c]: means in the same row with no common superscripts differ significantly ($p < 0.05$). [1]: Obtained from 36-day-old broilers. T1: 20% crude protein content, valine/lysine ratio (0.80) and arginine/lysine ratio (1.05) formulated according to current recommendations; T2: 18% crude protein content and valine/lysine ratio (0.80) and arginine/lysine ratio (1.05) formulated according to current recommendations; T3: 18% crude protein content, below-required valine/lysine ratio (0.70) and arginine/lysine ratio (1.05) formulated according to current recommendations; T4: 18% crude protein content, below-required arginine/lysine ratio (0.92) and valine/lysine ratio (0.80) formulated according to current recommendations.

2.4. Discussion

The serum uric N and productive traits were within the normal parameters obtained for broilers in the grower phase, and they agree with previous work [36]. Our data suggest that none of the diets offered with a low crude protein level (18% in T2 to T4) achieved the productive traits obtained with diet T1 (20% crude protein). Some authors indicate that establishing a minimum dietary crude protein content may not be necessary when proper amino acid ratios are implemented in diet formulation [37]. Moreover, research has shown that reductions in crude protein levels (below 19.5%) in broilers of similar ages can limit growth [38,39].

Animals fed diet T4 showed a significantly higher SUN compared with the rest of the animals in groups T1 to T3. Higher SUN could indicate major amino acid catabolism. Therefore, according to changes in SUN concentration, T4 would be the most unbalanced diet in terms of amino acids. These results agree with low performance data (weight and ADG) in T4.

Figure 1 shows there is high individual variability in the ADG and in the SUN content amongst animals, even for those within the same dietary treatment. This figure shows that the animals fed the highest protein content (diet T1) are mostly in the upper half (high growth rate), and that in the low-growth and high-SUN quadrant, there are mainly animals fed the diet with a low arginine to lysine ratio (T4).

As regards low-protein diets, animals fed diet T2 (with valine/lysine and arginine/lysine according to current recommendations) showed similar SUN levels to those fed diet T3 (with valine/lysine ratio below current recommendations), but lower SUN levels (−16%; $p < 0.05$) than those fed diet T4 (with arginine/lysine ratio below current recommendations). These results indicate that current recommendations of valine and arginine seem to be correctly determined, as a reduction in any of them has a clear negative effect on broiler growth performance. However, only a reduction in arginine (and not valine) increased amino acids catabolism. This could imply that when lysine and methionine are well-fitted, arginine would limit the protein use of the animals more than valine (in diets with low crude protein levels). It would be interesting to verify arginine levels, since it could be the third limiting amino acid when low-protein diets based on corn, wheat and soybean meal are used in broilers. Some authors have already stated the importance of arginine when the protein level is limited [40]. In addition, positive effects have been shown when elevated levels of arginine were supplied under these conditions [41,42]. Attia et al. [43] suggested that the response to the level of amino acid addition in low-protein diets can also vary according to bird strain and age.

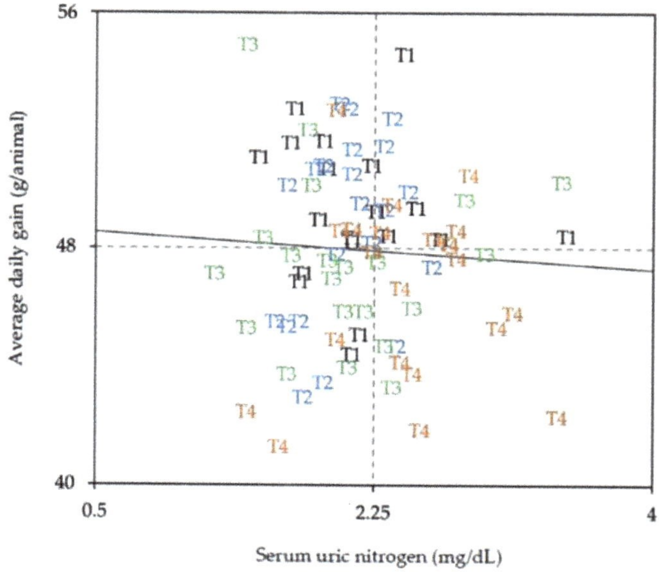

Figure 1. Relationship between animal's serum uric nitrogen (SUN), when fed the different experimental diets varying in crude protein, as well as valine and arginine to lysine ratios (T1 to T4), and individual average daily weight gain during the last period of the grower phase (day 28 to 35) in broilers (n = 21 animals per treatment). T1: 20% crude protein content, valine/lysine ratio (0.80) and arginine/lysine ratio (1.05) formulated according to current recommendations; T2: 18% crude protein content and valine/lysine ratio (0.80) and arginine/lysine ratio (1.05) formulated according to current recommendations; T3: 18% crude protein content, below-required valine/lysine ratio (0.70) and arginine/lysine ratio (1.05) formulated according to current recommendations; T4: 18% crude protein content, below-required arginine/lysine ratio (0.92) and valine/lysine ratio (0.80) formulated according to current recommendations.

Although more studies are necessary to establish the potential of SUN and other biomarkers (as glutamine or glutamate in the blood) to determine amino acid imbalances in broiler diets, this study highlights the interaction of nutrition with metabolic phenotype to achieve this goal.

3. Case Study 2: *De Novo* Protein Design, an Example of an Ideal Protein for 21-Day-Old Broilers

3.1. Background

The ideal protein concept is built on the idea that birds need specific amounts and ratios of amino acids to achieve their optimal performance and maximum growth [11]. Dietary amino acid concentrations should match needs for both maintenance and muscle accretion to effectively allow for the increased synthesis of white meat [44]. However, the utilization efficiency (including digestion and metabolism) of amino acids from feed ingredients is relatively low, resulting in high N losses in excreta [13].

Moreover, with current feed formulations, the right proportion of amino acids for each animal cannot be properly achieved for all amino acids at the same time without producing an excess of some amino acids to ensure others. In addition, protein digestibility depends not only on the molecular features of the protein, but also on the action of the enzymes (i.e., proteases) involved in the digestion process. In order to allow absorption by enterocytes in the small intestinal mucosa, proteins must be broken down into dipeptides, tripeptides or free amino acids. The specificity of enzymes (stomach pepsin, pancreatic trypsin and chymotrypsin, as well as intestinal mucosal carboxypeptidases and aminopeptidases), their enzyme to substrate ratio, as well as their different ways of cleaving peptide bonds will determine the final level of protein breakdown. The structural properties of proteins (secondary structure and β-conformations) may play a major role in resistance to denaturation and gastrointestinal digestion, as well [45].

Computational and protein engineering methods could be valuable tools to help design a protein sequence and structure that meets the needs of all amino acids (without excesses or defects), and which is fully digested and metabolized in broilers. The obtained protein could be synthesized and used in the future to contribute to this goal.

In recent decades, the use of certain synthetic amino acids has allowed us to better adjust diets and reduce their protein contents. Although we are far from being able to develop completely synthetic proteins that can cover these needs in a profitable way, the exponential development of biotechnology could get us close to that reality in the coming years. Therefore, knowledge of how to develop this technology is necessary.

This case study illustrates how *de novo* protein design can contribute to this aim. It describes a novel stepwise modeling approach to designing an ideal protein that can be completely digestible and usable in broilers from 0 to 21 days of age. It is an example of how precision nutrition strategies combining traditional nutrition with biotechnological, computational and protein engineering approaches can contribute to addressing precision poultry nutrition challenges in the future. The boundary conditions set in this case scenario include: (i) a primary protein sequence design containing the minimal amino acid quantities that can be fully digested by enzymes from the avian digestive system, and (ii) modeling the secondary and tertiary conformations of already-designed polypeptides.

3.2. Experimental Procedure

To this end, firstly, a literature review was conducted to define the requirements of amino acids for 21-day-old broilers, as regards the amino acid composition and ideal protein profile. Secondly, the protein digestion dynamics and functioning of the digestive system at the enzymatic level of chickens of that age were also reviewed. From these data, potential primary polypeptide sequences were designed. We chose the shortest protein sequence that fulfilled the following criteria: (i) fully meeting broiler requirements of all amino acids, while (ii) optimizing digestive enzyme functioning. Finally, we predicted its secondary and tertiary structure and its physicochemical properties using computational methods.

The *de novo* design of a protein that fully meets the requirements of broilers should be based on the net amino acid requirement (maintenance plus growth requirements) at each stage of the animal's life. Although there have been some attempts to obtain this information [46], it is not yet available. For this reason, the present work was based on the closest estimates of net requirements, which correspond to the true ileal digestible amino acids (i.e., corrected for basal and specific endogenous amino acid secretions).

The amino acid requirements are outlined in Table 3, based on the ideal amino acid profiles for broilers from 0 to 21 days with respect to lysine, as proposed by the North American Texas AM University [47]. These data were selected after comparing them with the available literature and recognized international nutritional guidelines for broilers (Canadian NRC [48], Spanish FEDNA [49], Dutch CVB [50] and Brazilian Tables for poultry and swine [51]). Wu's [47] dataset, containing a total of 108 amino acids (Table 3), was chosen to construct the "minimal ideal protein", because these recommendations are not far from the current recommendations for most of the amino acids provided by FEDNA [49] and NRC [48]. Moreover, it was the only one that provided recommendations for the 20 amino acids, and it was derived from true ileal digestible amino acid contents, accounting for the proportions of amino acids in the whole bodies of broilers.

Table 3. Amino acid requirements (expressed relative to lysine, lysine = 100) for chickens from 0 to 21 days used for protein modeling.

Amino Acid	Mw (g/mol)	Amino Acid/Lysine [1]	Molecules in the Sequence [2]
Alanine	89.09	102	6
Arginine	174.2	105	7
Asparagine	132.1	56	4
Aspartate	133.1	66	4
Cystein	121.2	32	2
Glutamate	147.1	178	11
Glutamine	146.2	128	8
Glycine	75.1	176	11
Histidine	155.2	35	2
Isoleucine	131.2	67	4
Leucine	131.2	109	7
Lysine	146.2	100	6
Methionine	149.2	40	3
Phenylalanine	165.2	60	4
Proline	115.1	184	12
Serine	105.1	69	4
Threonine	119.1	67	4
Tryptophan	204.2	16	1
Tyrosine	181.2	45	3
Valine	117.2	77	5
		Total number of amino acids	108
		Mw (g/mol) [3]	12095.1

Mw: molecular weight. [1]: Calculated from true ileal digestibility data from Wu [47]. [2]: As the amino acid that is required in the lowest relative proportion with respect to lysine is tryptophan at 0.16, the number of molecules in the sequence was calculated to have at least one representative of tryptophan in the protein sequence. The rest of the amino acids were proportionally calculated according to this value, as (amino acid/lysine)/0.16. [3]: Calculated as the sum of the individual amino acid's molecular weight × the number of molecules in the sequence, minus the molecular weight of the 107 peptide bonds (condensation reactions; 18 g/mol per bond) needed to generate the 108-amino acid sequence.

From the 108 amino acids described in Table 3, an initial protein sequence was generated using RandSeq (from the ExPASy online portal, SIB Bioinformatics Resource Portal). This online tool is frequently used to build randomly scrambled peptide libraries from a specific amino acid composition [52,53].

Using the random sequence obtained using RandSeq, several primary structures were designed using the Peptide Cutter software's information (ExPASy Bioinformatics Portal, Swiss Institute of Bioinformatics). The Peptide Cutter software considers the performance (activity and substrate specificity) of avian digestion enzymes in the sequence and maximizes the number of cleavages by enzymes in the linear polypeptide chain. The choice of enzymes was based on the work of Recoules et al. [54], using in vivo data on the digestion of plant proteins in broilers. Pepsin, trypsin, chymotrypsin, elastase, prolidase, carboxypeptidase A and B and aminopeptidase were studied.

The final sequences obtained were subjected to a manual refinement step to increase their potential digestibility. In other words, we increased the number of free amino acids in the final sequence by adding extra specific amino acids that would break the remaining dipeptides. Such extra amino acids were chosen following two criteria: being the target amino acid of various avian digestive enzymes and having been rounded down in the proposed minimum ideal protein initial sequence.

After this step, its secondary and tertiary structures were predicted using two online servers: i-TASSER (iterative threading assembly refinement, a hierarchical protocol for the structural and functional prediction of amino acid sequences [55]) and QUARK (based on ab initio folding, the construction of protein structures by fragment assembly from unrelated proteins [56]). Both software were used to predict the folding of sequences.

3.3. Results

Figure 2 shows the optimal initial primary amino acid sequence derived from the information in Table 3, and the protein digestion dynamics data from enzyme affinities (Round 3, Figure 1). The round 3 sequence was the most digestible sequence based on the action of the chicken digestive enzymes, because it led to a high number of free amino acids after digestion—only four dipeptides and 100 free-amino acids (7.4 and 92.6% of the total amino acids in the sequence, respectively).

Round 3: M-E-P-F-V-N-P-F-A-E-P-L-Q-K-T-R-H-N-I-Q-P-L-V-G-G-A-V-E-P-L-N-K-E-W-G-Q-Y-Q-R-G-E-P-F-Q-Q-R-H-G-S-R-E-K-G-Q-
I-N-P-L-M-E-P-L-S-R-T-I-Q-I-D-R-Q-I-A-S-K-D-K-A-D-W-D-P-L-A-T-K-G-S-R-V-E-Y-G-M-V-C-K-G-E-P-F-E-R-A-G-C-P-L-T-P-Y-E

Round 3.1: M-G-A-E-Y-S-K-A-Q-P-L-Q-I-S-P-F-Q-R-M-S-R-G-E-P-F-E-P-W-N-K-Y-A-E-P-L-D-P-L-T-R-M-G-V-G-V-Q-P-F-G-Q-K-G-N-R-
E-K-Q-R-C-K-A-N-I-T-I-N-P-F-G-E-K-G-D-P-L-A-E-R-H-D-I-G-D-P-L-E-T-I-V-H-G-G-S-K-E-R-V-E-R-A-C-P-L-Q-Y-T-P-L-E-W-V-Q

Round 3.2: M-V-E-P-L-E-R-G-D-P-L-Q-I-T-P-F-V-E-K-V-N-P-L-T-K-N-P-W-Q-P-F-E-I-G-H-G-G-G-D-R-T-K-Q-Y-E-R-G-S-K-E-R-G-A-D-
I-E-K-M-V-S-K-A-Q-P-L-E-I-S-I-G-E-R-A-A-Q-Y-Q-W-M-G-S-R-V-G-D-P-L-Q-P-L-N-P-F-T-P-F-A-E-K-G-Q-R-C-Y-A-N-P-L-E-R-H-C

Round 3.3: M-Q-R-G-Q-R-E-R-A-G-T-I-T-I-C-P-F-E-P-F-T-W-V-D-K-G-N-R-V-V-S-P-F-G-E-R-Q-I-A-Q-P-L-G-E-P-L-C-Y-E-K-G-S-P-L-
G-E-K-A-N-K-V-E-P-L-G-V-M-S-K-T-K-H-A-D-P-F-E-Y-Q-R-E-R-A-N-P-L-E-I-G-G-S-P-L-D-R-N-I-D-Y-Q-P-W-E-P-L-M-Q-K-G-H-A-Q

Round 3.4: M-G-Q-R-A-E-P-F-E-R-G-T-P-L-Q-W-G-T-P-F-S-I-G-H-A-Q-K-S-I-E-P-L-A-G-N-I-N-K-D-G-E-I-Q-P-L-V-G-E-K-G-T-K-E-R-
G-E-P-L-E-P-F-Q-Y-S-P-L-N-P-L-C-K-V-N-P-F-V-R-D-R-E-R-C-Y-H-A-M-D-K-Q-R-D-R-A-V-T-Y-G-E-P-L-S-I-M-G-Q-P-W-V-E-K-A-Q

Figure 2. Original 108-amino acid sequence (Round 3) and refined sequence with 112 amino acids modeled for complete digestion (Round 3.1, Round 3.2, Round 3.3 and Round 3.4). One-letter amino acid code: A—alanine, C—cysteine, D—aspartic acid, E—glutamic acid, F—phenylalanine, G—glycine, H—histidine, I—isoleucine, K—lysine, L—leucine, M—methionine, N—asparagine, P—proline, Q—glutamine, R—arginine, S—serine, T—threonine, V—valine, W—tryptophan, Y—tyrosine.

Rounds 3.1 to 3.4 (Figure 2) were extra sequences generated from the original sequence (Round 3) during manual refinement. These extra sequences (Rounds 3.1 to 3.4) were designed to break the four remaining dipeptides following complete digestion. To this end, four extra amino acids were included in the composition (a total of 112 amino acids), with the following considerations: (i) prioritizing those amino acids that were a frequent target for digestive enzymes in chickens (arginine, isoleucine, leucine, lysine, phenylalanine, tryptophan and tyrosine); (ii) promoting isoleucine addition, given that it is rounded down to avoid shortage; (iii) giving special attention to lysine, due to its roles as the first limiting and the reference amino acid, making it worthwhile to ensure its minimal requirement is met; (iv) adding arginine and tryptophan, due to their relevance as first limiting amino acids; (v) avoiding cysteine (due to the risk of disulphide bridges), which reduces digestive enzymes' efficiency. The addition of the four extra amino acids resulted in increases in the amounts of isoleucine by 25.0%, lysine by 16.7%, arginine by 14.3% and tryptophan by 100.0%.

Finally, secondary and tertiary protein structures were determined and tested to evaluate the quality and reliability criteria of the structural models obtained with the different protein sequences, using I-TASSER and QUARK software. Figure 3 presents the Round 3.3 sequence, which was the most reliable and highest-quality model, as indicated by its C-score (accuracy ranging from −5 to 2, increasing with high confidence) and TM-score (similarity to native structures, with a TM-score > 0.5, similar topology, and <0.3 random similarity). The C-score and TM-score values in Round 3.3 were higher than those of the other sequences determined using I-TASSER (on average, C-score −1.08 vs. −3.45, and TM-score 0.58 vs. 0.34). Theoretically, this implies higher reliability when predicting the actual 3D protein conformation. A complete descriptionvs. of all sequence quality and reliability criteria has been given by Lledó [57].

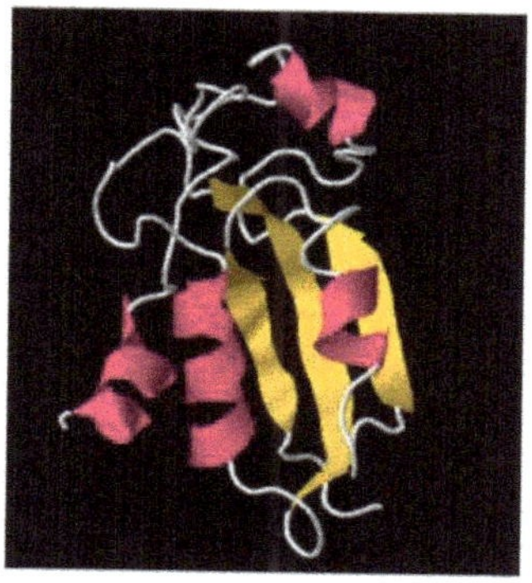

(**a**)

```
                       20                  40                  60                  80                  100
                        |                   |                   |                   |                   |
Sequence   MQRGQRERAGTITICPFEPFTWVDKGNRVVSPFGERQIAQPLGEPLCYEKGSPLGEKANKVEPLGVMSKTKHADPFEYQRERANPLEIGGSPLDRNIDYQPWEPLMQKGHAQ
Prediction CCCCCCCCCCSSSSCCCCCCSSSCCCCSSSCCCCHHHHHCCCCCCCSSCCCCCCCCCCCCCCCCSSSCCCCCCCHHHHHHHHCCCCCCCCCCCCCCCCCCCHHHHHCCCCC
Conf.Score 9755211166158626777055606985636631233303257854403588502113453543124245458806666630784112787555656777562777633579
           H:Helix; S:Strand; C:Coil
```

(**b**)

Figure 3. Predicted secondary and tertiary structure of sequence Round 3.3 by I-TASSER. (**a**) Three-dimensional tertiary structure cartoon model. In pink α-helices, in yellow β-sheets and in white coil regions. (**b**) Secondary predicted structure. H: α-helices; S: β-sheets; C: coil regions based on *in silico* digestion of initial designed sequences for primary structure. One letter amino acid code: A—alanine, C—cysteine, D—aspartic acid, E—glutamic acid, F—phenylalanine, G—glycine, H—histidine, I—isoleucine, K—lysine, L—leucine, M—methionine, N—asparagine, P—proline, Q—glutamine, R—arginine, S—serine, T—threonine, V—valine, W—tryptophan, Y—tyrosine.

Figure 3 shows the overall sequence covered by α-helices, β-sheets and random coil regions, derived from the i-TASSER models of Round 3.3. The secondary structure of the protein simulated in Round 3.3 showed the highest amount of β structures (13%) and the least α-helices (16%) among all candidates, and therefore it may be the least digestible protein. Following the same criteria, the secondary structure of the protein modelled in Round 3.1 (Figure 4) contained the lowest percentage of β-sheets (2%), and simultaneously the highest number of α-helices (41%), among all the models. Regarding the number of coil regions, as their conformational prediction is more intricate, these regions could display more unexpected folds. Therefore, defined structures and α-helices are preferable. The structure of the protein simulated in Round 3.1 presented one of the lowest percentages of coil regions (on average, 57% vs. 70%).

Therefore, based on the structural motifs, the protein with the highest number of α-helices and the lowest number of β-sheet was that in Round 3.1. Moreover, it was ranked second in terms of reliability on the basis of its C-score and TM-score, showing a more acceptable quality level (−1.99 and 0.48, respectively) with respect to Round 3.3 (−1.08 and 0.58, respectively). Figure 4 shows the structural properties of the Round 3.1 protein modeled using different software.

3.4. Discussion

The efficiency of the use of ingested dietary protein by broilers depends on the digestibility and balance of the amino acid content relative to the animal's requirements. Increasing the crude protein content has been proven to entail negative effects in chicken health, and in environmental and production costs [58]. On the other hand, low-crude protein level diets with the addition of crystalline amino acid do not constitute a complete solution, because this can reduce chicken growth performance.

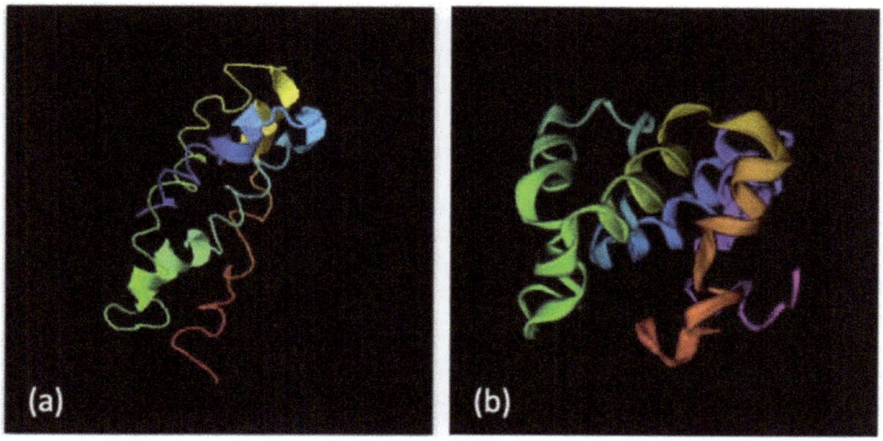

Figure 4. Protein 3D structure model of Round 3.1 using (**a**) the I-TASSER model and (**b**) the QUARK model.

The "perfect" diet, in terms of protein supply, could involve feeds with low level inclusions of highly purified and digestible proteins. This work has outlined a novel approach, combining the structural digestibility, quality, and reliability criteria of the predicted model, resulting in a valid protein design that will help to achieve this goal.

The resulting protein model had a minimal size, with 112 amino acids. This size seems adequate to produce the ideal protein, as it is the closest to the ideal amino acid profile of broilers. Low-molecular weight proteins are more easily produced and secreted, and are less likely to interact with the host's metabolism. This could be advantageous in terms of future biological synthesis and industrial production.

In terms of protein digestibility and solubility, it is crucial to consider the occurrence of two main structural patterns: α-helices and β-sheets. Carbonaro et al. [45] studied in vitro the structure–digestibility relationship of different proteins of animal and plant origin, and quantified the different structural motifs. Their results, consistent with other experiments [59,60], showed a reduction in hydrolysis degree that was inversely proportional to the number of β-sheets. The main explanation for this lies in the hydrophobic character of these structures, which promotes aggregation and protein–protein interactions.

Obtaining an accurate model for secondary and tertiary structures is essential, since these structures are closely related to proteins' physical (solubility, aggregation and secretory ability) and functional characteristics. Protein solubility is very relevant to production processes, as it mainly affects cell excretion and downstream recovery processes, given that it is directly related to the aggregation phenomena. Moreover, secondary structural motifs can also affect peptide solubility. Solubility has been shown to increase with the ratio of α-helices to β-sheets in in vitro experiments [61].

Besides this, secondary and tertiary structural modeling can give clues about protein functional features and behaviors. The potential of the protein prototype is also strongly determined by the existence of protein templates in protein data banks with significant sequence similarity to the problem sequence. In other words, sequences without equivalents in protein data banks will be more difficult to model, and the result will be based on less evidence, leading to less reliable results overall. However, a unique protein, whose structure greatly differs from any known homologous proteins, could be advantageous. Firstly, it could decrease the mimic phenomenon, which could lead to problems in the host used for future production (bacteria, yeast, or any other chosen organism). Secondly, it is more probable that, if it does not belong to any protein family, it will not have any relevant function itself.

Through the procedure followed in this work, we obtained a prototype that meets most of the conditions that a synthetic protein should have, such as being completely digestible, not generating an excess supply of amino acids (since it is ideally adjusted to the requirements of the targeted animal), and therefore coming as close as possible to the concept of an ideal protein. In any case, more studies are necessary to improve the definition of this protein prototype that will consider the characteristics of the potential hosts, before carrying out pilot tests aimed at biosynthesis. Current protein yields using plant or microbial fermentation synthesis are still insufficient to produce a fully viable synthetic protein source for poultry. However, the efficiency of these processes is rapidly improving, and new developments in bioprocesses are emerging [62], as are innovative procedures, such as cell-free protein synthesis [63]. The use of these novel protein synthesis methods could contribute to making this procedure a reality in the near future.

4. Future Precision Nutrition Needs to Improve Protein Nutrition in Broilers

Precision nutrition is an essential part of precision livestock farming, as both pursue a common goal: enhancing farm profitability, efficiency, and sustainability [64]. In our work, potential tools for the development of future precision nutrition strategies that will improve amino acid utilization in broilers are presented and discussed. The tools presented herein include biotechnological, metabolomic, computational and protein engineering approaches, and they are summarized in Figure 5, focusing on metabolic phenotyping and the identification of individual variability using biomarkers (Case study 1), and accurate feed matrix formulation and ingredient design (partial or complete) through *de novo* protein development (Case study 2).

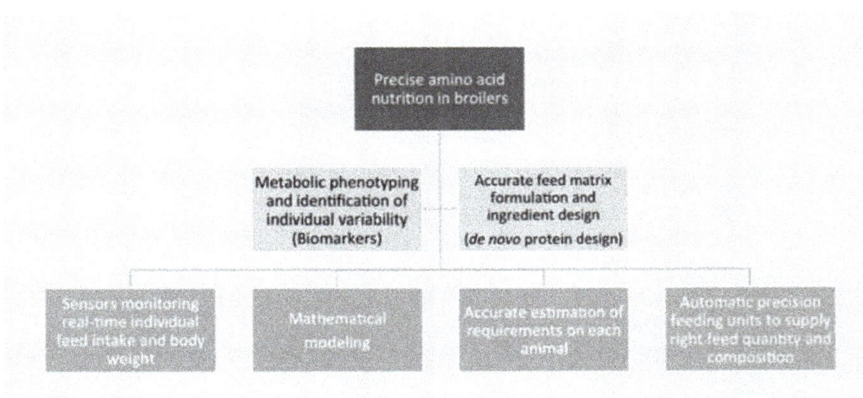

Figure 5. Integrated framework scheme for precise amino acid nutrition in broiler farming, combining nutritional strategies and feeding technologies.

The precision nutrition strategies presented in this work differ from precision feeding technologies. The latter relies on monitoring the amount and composition of the feed delivered to broilers, individually or in groups, as well as their feeding behavior and intake [4]. Precision feeding technologies require the development of sensors and automatic measuring devices, as well as sophisticated feeding units capable of providing the required amount and composition of feeds according to specific production objectives, based on growth models derived from computational methods, i.e., customized diets [65].

Therefore, the on-farm application of precision feeding technologies in broilers is still limited, and there are few examples of the practical use of such technologies. For instance, in poultry, individual monitoring is complex, and it has only been implemented in broiler breeders using feeding cubicles, based on using bird radio frequency identification (RFID), weighing in the cubicle, and a target growth curve updated in real-time [66]. Although RFID systems can accurately detect and record the feeding behaviors of individual broilers in group settings automatically [67], their field implementation is challenging, costly and complex. Furthermore, the estimation of individual nutritional requirements in real-time is not entirely feasible, and must be based on mathematical models and theoretical estimations.

In contrast, the precision nutrition strategies addressed in this work are designed to equal the dietary nutrient supply to the nutrient requirements of the animals, particularly focusing on amino acids. There is still a need to adjust the exact combination of indispensable amino acids so that they exactly match the animals' requirements for protein accretion and maintenance, with no deficiencies and no excesses using the ideal protein profile [11]; but also to come up with valid tools that can give real feedback using animal-based biomarkers. These strategies are necessary prerequisites that must be implemented in future automated and tailored feeding technologies. In fact, as Moss et al. [4] stated, the implementation of precision nutrition relies on the ability of the poultry industry to employ precision feeding within its operations, and therefore, precise nutrition strategies must be combined with precise feeding technologies.

The tools addressed in the present case studies are therefore key to formulating an integrated framework for precise amino acid nutrition in broilers. Figure 5 illustrates how biomarkers and *de novo* protein design can fit into this scheme, as indispensable to the precise livestock farming matrix that combines feeding technologies (including sensors and automatic feeding units), modeling at individual or group levels, and nutritional strategies in an interdisciplinary approach.

5. Conclusions

Specific biomarkers, such as SUN, can be useful as rapid indicators of individual or group amino acid deficiencies or imbalances. Their practical implementation on-farm, as well as in nutritional research studies, could contribute to the achievement of both precise diet formulation and the determination of nutrient requirements in broilers. These types of biomarkers could be a suitable tool for closing the gap between models and farm conditions, as they could be used as an indicator of how the modelled theoretical or estimated requirements need to be adjusted in farm conditions. Broilers in different farm settings can be exposed to stressors (environmental, social and nutritional, amongst others). Individual metabolic phenotyping using biomarkers could contribute to optimizing nutrient use, reducing safety margins, and preventing nutrient over-feeding.

Furthermore, the possibility of designing a specific protein that can minimize N losses, and maximize its digestibility and metabolic use, was discussed in this work. We developed a stepwise approach to designing an ideal protein that can be completely digestible and usable in broilers from 0 to 21 days of age. The procedure presented in this work is promising for the initial design of synthetic oligopeptides that could be used in a similar way to current synthetic amino acids, and ultimately for the synthesis of proteins that meet the needs of animals exactly. Its application could help us to precisely match nutrient supply with the nutrient requirements of animals without excesses, thus minimizing losses.

In conclusion, both tools presented herein can open up new opportunities in the context of future broiler nutrition and precision farming, helping us to achieve more efficient, resilient, and sustainable production. This information can help us to determine the exact ratio of amino acids that will improve the efficiency of the use of N by poultry.

Author Contributions: Conceptualization, M.C.-L., P.J.M.-G. and J.J.P.; methodology, M.C.-L., P.J.M.-G., C.L., A.C. and J.J.P.; validation, A.C. and J.J.P.; formal analysis, P.J.M.-G. and C.L.; investigation, M.C.-L., P.J.M.-G., C.L., A.C. and J.J.P.; resources, A.C. and J.J.P.; data curation, M.C.-L. and J.J.P.; writing—original draft preparation, M.C.-L., P.J.M.-G. and J.J.P.; writing—review and editing, M.C.-L. and J.J.P.; supervision, M.C.-L. All authors have read and agreed to the published version of the manuscript.

Funding: This research has not received any specific grant from any public funding agency.

Institutional Review Board Statement: The protocol for the study on animals was reviewed and approved by the Ethics Committee of the Universitat Politècnica de València (code 2017/VSC/PEA/000166). All the experiments were carried out at the Animal Experimentation Centre (code: ES121040000044) in accordance with the guidelines and regulations established in Directive 2010/63/EU EEC and Royal Decree 53/2013 on the protection of animals used for scientific purposes.

Informed Consent Statement: Not applicable.

Data Availability Statement: Data is contained within the article.

Acknowledgments: The authors wish to thank the technical staff at the experimental farm at Centro de Investigación de Tecnología Animal (CITA-IVIA), as well as the technical staff at the Animal Nutrition Laboratory (Universitat Politècnica de València, UPV).

Conflicts of Interest: The authors declare no conflict of interest.

References

1. Sibbald, I.R. A Bioassay for Available Amino Acids and True Metabolizable Energy in Feedstuffs. *Poult. Sci.* **1979**, *58*, 668–673. [CrossRef]
2. Maharjan, P.; Liang, Y. Precision Livestock Farming: The Opportunities in Poultry Sector. *J. Agric. Sci. Technol. A* **2020**, *10*, 45–53. [CrossRef]
3. Banhazi, T.; Lehr, H.; Black, J.; Crabtree, H.; Schofield, C.; Tscharke, M.; Berckmans, D. Precision Livestock Farming: An International Review of Scientific and Commercial Aspects. *Int. J. Agric. Biol. Eng.* **2012**, *5*, 1. [CrossRef]
4. Moss, A.F.; Chrystal, P.V.; Cadogan, D.J.; Wilkinson, S.J.; Crowley, T.M.; Choct, M. Precision Feeding and Precision Nutri-tion: A Paradigm Shift in Broiler Feed Formulation? *Anim. Biosci.* **2021**, *34*, 354–362. [CrossRef] [PubMed]
5. Warren, W.A.; Emmert, J.L. Efficacy of Phase-Feeding in Supporting Growth Performance of Broiler Chicks during the Star-ter and Finisher Phases. *Poult. Sci.* **2000**, *79*, 764–770. [CrossRef]
6. Hauschild, L.; Bueno, C.F.D.; Remus, A.; Gobi, J.D.P.; Isola, R.D.G.; Sakomura, N.K. Multiphase Feeding Program for Broilers Can Replace Traditional System. *Sci. Agric.* **2015**, *72*, 210–214. [CrossRef]
7. Currie, R.; Maclsaac, J.; Rathgeber, B. Increasing the Number of Phases Fed to Broiler Chickens by Blending Diets. *Poult. Sci.* **2006**, *85*, 52.
8. Pomar, C.; Remus, A. Precision Pig Feeding: A Breakthrough toward Sustainability. *Anim. Front.* **2019**, *9*, 52–59. [CrossRef]
9. Halachmi, I.; Guarino, M. Editorial: Precision Livestock Farming: A 'per Animal' Approach Using Advanced Monitoring Technologies. *Animal* **2016**, *10*, 1482–1483. [CrossRef]
10. Liu, S.Y.; Macelline, S.P.; Chrystal, P.V.; Selle, P.H. Progress towards Reduced-Crude Protein Diets for Broiler Chickens and Sustainable Chicken-Meat Production. *J. Anim. Sci. Biotechnol.* **2021**, *12*, 20. [CrossRef]
11. Emmert, J.L.; Baker, D.H. Use of the Ideal Protein Concept for Precision Formulation of Amino Acid Levels in Broiler Diets. *J. Appl. Poult. Res.* **1997**, *6*, 462–470. [CrossRef]
12. Ferket, P.R.; van Heugten, E.; van Kempen, T.A.T.G.; Angel, R. Nutritional Strategies to Reduce Environmental Emissions from Nonruminants. *J. Anim. Sci.* **2002**, *80*, E168–E182. [CrossRef]
13. Powers, W.; Angel, R. A Review of the Capacity for Nutritional Strategies to Address Environmental Challenges in Poultry Production. *Poult. Sci.* **2008**, *87*, 1929–1938. [CrossRef] [PubMed]
14. Macfarlane, G.T.; Macfarlane, S. Bacteria, Colonic Fermentation, and Gastrointestinal Health. *J. AOAC Int.* **2012**, *95*, 50–60. [CrossRef] [PubMed]
15. Gilbert, M.S.; Ijssennagger, N.; Kies, A.K.; van Mil, S.W.C. Protein Fermentation in the Gut; Implications for Intestinal Dys-function in Humans, Pigs, and Poultry. *Am. J. Physiol. Gastrointest. Liver Physiol.* **2018**, *315*, G159–G170. [CrossRef]
16. Celi, P.; Cowieson, A.J.; Fru-Nji, F.; Steinert, R.E.; Kluenter, A.-M.; Verlhac, V. Gastrointestinal Functionality in Animal Nu-trition and Health: New Opportunities for Sustainable Animal Production. *Anim. Feed Sci. Technol.* **2017**, *234*, 88–100. [CrossRef]
17. Bindari, Y.R.; Gerber, P.F. Centennial Review: Factors Affecting the Chicken Gastrointestinal Microbial Composition and Their Association with Gut Health and Productive Performance. *Poult. Sci.* **2022**, *101*, 101612. [CrossRef]
18. McDonald, P.; Edwards, R.A.; Greenhalgh, J.F.D.; Morgan, C.A.; Sinclair, L.A.; Wilkinson, L.G. *Animal Nutrition*, 7th ed.; Pearson, Prentice Hall: Hoboken, NJ, USA, 2011.
19. Steed, J.R.; Romero-Sanchez, H.; Han, Y.; Page, G.I.; Davis, A.J. Validation of NutriOpt Dietary Formulation Strategies on Broiler Growth and Economic Performance. *J. Appl. Poult. Res.* **2020**, *29*, 314–327. [CrossRef]
20. Halas, V. Growth Models and Their Application in Precision Feeding of Monogastric Farm Animals. *Acta Fytotechn. Zootechn.* **2020**, *23*, 258–264. [CrossRef]
21. Coma, J.; Zimmerman, D.R.; Carrion, D. Lysine Requirement of the Lactating Sow Determined by Using Plasma Urea Nitro-gen as a Rapid Response Criterion. *J. Anim. Sci.* **1996**, *74*, 1056. [CrossRef]

22. Nieto, R.; Barea, R.; Lara, L.; Palma-Granados, P.; Aguilera, J.F. Lysine Requirement Relative to Total Dietary Protein for Optimum Performance and Carcass Protein Deposition of Iberian Piglets. *Anim. Feed Sci. Technol.* **2015**, *206*, 48–56. [CrossRef]
23. Roth-Maier, D.A.; Ott, H.; Roth, F.X.; Paulicks, B.R. Effects of the Level of Dietary Valine Supply on Amino Acids and Urea Concentration in Milk and Blood Plasma of Lactating Sows: Valine for Lactating Sows. *J. Anim. Physiol. Anim. Nutr.* **2004**, *88*, 39–45. [CrossRef] [PubMed]
24. Marín-García, P.J.; López-Luján, M.D.C.; Ródenas, L.; Martínez-Paredes, E.; Blas, E.; Pascual, J.J. Plasma Urea Nitrogen as an Indicator of Amino Acid Imbalance in Rabbit Diets. *World Rabbit Sci.* **2020**, *28*, 63. [CrossRef]
25. Donsbough, A.L.; Powell, S.; Waguespack, A.; Bidner, T.D.; Southern, L.L. Uric Acid, Urea, and Ammonia Concentrations in Serum and Uric Acid Concentration in Excreta as Indicators of Amino Acid Utilization in Diets for Broilers. *Poult. Sci.* **2010**, *89*, 287–294. [CrossRef]
26. Ravindran, V.; Bryden, W.L. Amino Acid Availability in Poultry—In Vitro and in Vivo Measurements. *Aust. J. Agric. Res.* **1999**, *50*, 889. [CrossRef]
27. Vieira, S.L.; Lemme, A.; Goldenberg, D.B.; Brugalli, I. Responses of Growing Broilers to Diets with Increased Sulfur Amino Acids to Lysine Ratios at Two Dietary Protein Levels. *Poult. Sci.* **2004**, *83*, 1307–1313. [CrossRef]
28. Han, Y.; Baker, D.H. Digestible Lysine Requirement of Male and Female Broiler Chicks During the Period Three to Six Weeks Posthatching. *Poult. Sci.* **1994**, *73*, 1739–1745. [CrossRef]
29. Han, Y.; Baker, D.H. Effects of Sex, Heat Stress, Body Weight, and Genetic Strain on the Dietary Lysine Requirement of Broi-ler Chicks. *Poult. Sci.* **1993**, *72*, 701–708. [CrossRef]
30. Han, Y.; Baker, D.H. Lysine Requirements of Fast- and Slow-Growing Broiler Chicks. *Poult. Sci.* **1991**, *70*, 2108–2114. [CrossRef]
31. Samadi; Liebert, F. Modelling the Optimal Lysine to Threonine Ratio in Growing Chickens Depending on Age and Efficiency of Dietary Amino Acid Utilisation. *Br. Poult. Sci.* **2008**, *49*, 45–54. [CrossRef]
32. Dozier, W.A.; Payne, R.L. Digestible Lysine Requirements of Female Broilers from 1 to 15 Days of Age. *J. Appl. Poult. Res.* **2012**, *21*, 348–357. [CrossRef]
33. Agostini, P.S.; Santos, R.R.; Khan, D.R.; Siebert, D.; van der Aar, P. The Optimum Valine: Lysine Ratios on Performance and Carcass Traits of Male Broilers Based on Different Regression Approaches. *Poult. Sci.* **2019**, *98*, 1310–1320. [CrossRef] [PubMed]
34. Zampiga, M.; Soglia, F.; Petracci, M.; Meluzzi, A.; Sirri, F. Effect of Different Arginine-to-Lysine Ratios in Broiler Chicken Diets on the Occurrence of Breast Myopathies and Meat Quality Attributes. *Poult. Sci.* **2019**, *98*, 2691–2697. [CrossRef] [PubMed]
35. Belloir, P.; Méda, B.; Lambert, W.; Corrent, E.; Juin, H.; Lessire, M.; Tesseraud, S. Reducing the CP Content in Broiler Feeds: Impact on Animal Performance, Meat Quality and Nitrogen Utilization. *Animal* **2017**, *11*, 1881–1889. [CrossRef]
36. Sigolo, S.; Deldar, E.; Seidavi, A.; Bouyeh, M.; Gallo, A.; Prandini, A. Effects of Dietary Surpluses of Methionine and Lysine on Growth Performance, Blood Serum Parameters, Immune Responses, and Carcass Traits of Broilers. *J. Appl. Anim. Res.* **2019**, *47*, 146–153. [CrossRef]
37. Kriseldi, R.; Tillman, P.B.; Jiang, Z.; Dozier, W.A. Effects of Feeding Reduced Crude Protein Diets on Growth Performance, Nitrogen Excretion, and Plasma Uric Acid Concentration of Broiler Chicks during the Starter Period. *Poult. Sci.* **2018**, *97*, 1614–1626. [CrossRef]
38. Berres, J.; Vieira, S.L.; Kidd, M.T.; Taschetto, D.; Freitas, D.M.; Barros, R.; Nogueira, E.T. Supplementing L-Valine and L-Isoleucine in Low-Protein Corn and Soybean Meal All-Vegetable Diets for Broilers. *J. Appl. Poult. Res.* **2010**, *19*, 373–379. [CrossRef]
39. Yuan, J.; Karimi, A.; Zornes, S.; Goodgame, S.; Mussini, F.; Lu, C.; Waldroup, P.W. Evaluation of the Role of Glycine in Low-Protein Amino Acid-Supplemented Diets. *J. Appl. Poult. Res.* **2012**, *21*, 726–737. [CrossRef]
40. Dilger, R.N.; Bryant-Angeloni, K.; Payne, R.L.; Lemme, A.; Parsons, C.M. Dietary Guanidino Acetic Acid Is an Efficacious Replacement for Arginine for Young Chicks. *Poult. Sci.* **2013**, *92*, 171–177. [CrossRef]
41. Jahanian, R. Immunological Responses as Affected by Dietary Protein and Arginine Concentrations in Starting Broiler Chicks. *Poult. Sci.* **2009**, *88*, 1818–1824. [CrossRef]
42. Mejia, L.; Zumwalt, C.D.; Tillman, P.B.; Shirley, R.B.; Corzo, A. Ratio Needs of Arginine Relative to Lysine of Male Broilers from 28 to 42 Days of Age during a Constant, Elevated Environmental Temperature Regimen. *J. Appl. Poult. Res.* **2012**, *21*, 305–310. [CrossRef]
43. Attia, Y.A.; Bovera, F.; Wang, J.; Al-Harthi, M.A.; Kim, W.K. Multiple Amino Acid Supplementations to Low-Protein Diets: Effect on Performance, Carcass Yield, Meat Quality and Nitrogen Excretion of Finishing Broilers under Hot Climate Conditions. *Animal* **2020**, *10*, 973. [CrossRef] [PubMed]
44. Vieira, S.L.; Angel, C.R. Optimizing Broiler Performance Using Different Amino Acid Density Diets: What Are the Limits? *J. Appl. Poult. Res.* **2012**, *21*, 149–155. [CrossRef]
45. Carbonaro, M.; Maselli, P.; Nucara, A. Relationship between Digestibility and Secondary Structure of Raw and Thermally Treated Legume Proteins: A Fourier Transform Infrared (FT-IR) Spectroscopic Study. *Amino Acids* **2012**, *43*, 911–921. [CrossRef]
46. Esteve, C. Towards Precision Poultry Feeding: Amino Acid Losses in the Body of Broiler Chickens Fed a Nitrogen-Free Diet. Master's Thesis, Universitat Politècnica de València, Valencia, Spain, 2021. Available online: http://hdl.handle.net/10251/174720 (accessed on 28 February 2022).
47. Wu, G. Dietary Requirements of Synthesizable Amino Acids by Animals: A Paradigm Shift in Protein Nutrition. *J. Anim. Sci. Biotechnol.* **2014**, *5*, 34. [CrossRef]
48. National Research Council (NRC). *Nutrient Requirements of Poultry*, 9th ed.; National Academy Press: Washington, DC, USA, 1994.

49. Santomá, G.; Mateos, G.G. *Necesidades Nutricionales En Avicultura, Normas FEDNA*, 2nd ed.; Fundación Española para el Desarrollo de la Nutrición Animal: Madrid, Spain, 2018.
50. Veeroederbureau, C. *Table Booklet Feeding of Poultry: Feeding Standards, Feeding Advices and Nutritional Values of Feed Ingredients*; CVB Series; Federatie Nederlandse Diervoederketen: Zoetermeer, The Netherlands, 2018.
51. Rostagno, H.S.; Texeira Albino, L.S.; Hannas, M.I.; Lopes Donzele, J.; Sakomura, N.; Perazzo, F.G.; Oliveira Brito, C. *Tablas Brasileñas Para Aves y Cerdos*, 4th ed.; Departamento de Zootecnia: Viçosa, Brazil, 2017.
52. Grishaeva, T.M.; Bogdanov, Y.F. On the Origin of Synaptonemal Complex Proteins. Search for Related Proteins in Proteo-mes of Algae, Lower Fungi, Mosses, and Protozoa. *Russ. J. Genet. Appl. Res.* **2013**, *3*, 481–486. [CrossRef]
53. Singh, M.; Kumar, V.; Sikka, K.; Thakur, R.; Harioudh, M.K.; Mishra, D.P.; Ghosh, J.K.; Siddiqi, M.I. Computational Design of Biologically Active Anticancer Peptides and Their Interactions with Heterogeneous POPC/POPS Lipid Membranes. *J. Chem. Inf. Model.* **2020**, *60*, 332–341. [CrossRef]
54. Recoules, E.; Sabboh-Jourdan, H.; Narcy, A.; Lessire, M.; Harichaux, G.; Labas, V.; Duclos, M.J.; Réhault-Godbert, S. Exploring the in Vivo Digestion of Plant Proteins in Broiler Chickens. *Poult. Sci.* **2017**, *96*, 1735–1747. [CrossRef]
55. Yang, J.; Zhang, Y. Protein Structure and Function Prediction Using I-TASSER. *Curr. Protoc. Bioinform.* **2015**, *52*, 5–8. [CrossRef]
56. Zhang, W.; Yang, J.; He, B.; Walker, S.E.; Zhang, H.; Govindarajoo, B.; Virtanen, J.; Xue, Z.; Shen, H.-B.; Zhang, Y. Integration of QUARK and I-TASSER for Ab Initio Protein Structure Prediction in CASP11: Ab Initio Structure Prediction in CASP11. *Proteins* **2016**, *84*, 76–86. [CrossRef]
57. Lledó, C. De Novo Design of an Ideal Protein for Feeding Meat Chickens from 0 to 21 Days. Bachelor's Thesis, Universitat Politècnica de València, Valencia, Spain, 2020. Available online: http://hdl.handle.net/10251/149623 (accessed on 28 February 2022).
58. Esmail, S.H. Understanding Protein Requirements. 2016. Available online: https://www.poultryworld.net/Nutrition/Articles/2016/11/Understanding-protein-requirements-2914798W/ (accessed on 20 May 2020).
59. Gabriel, I.; Quillien, L.; Cassecuelle, F.; Marget, P.; Juin, H.; Lessire, M.; Sève, B.; Duc, G.; Burstin, J. Variation in Seed Protein Digestion of Different Pea (Pisum Sativum L.) Genotypes by Cecectomized Broiler Chickens: 2. Relation between in Vivo Protein Digestibility and Pea Seed Characteristics, and Identification of Resistant Pea Polypeptides. *Livest. Sci.* **2008**, *113*, 262–273. [CrossRef]
60. Yang, Y.; Wang, Z.; Wang, R.; Sui, X.; Qi, B.; Han, F.; Li, Y.; Jiang, L. Secondary Structure and Subunit Composition of Soy Protein In Vitro Digested by Pepsin and Its Relation with Digestibility. *Biomed. Res. Int.* **2016**, *2016*, 5498639. [CrossRef] [PubMed]
61. Bai, M.; Qin, G.; Sun, Z.; Long, G. Relationship between Molecular Structure Characteristics of Feed Proteins and Protein in Vitro Digestibility and Solubility. Asian Australas. *J. Anim. Sci.* **2015**, *29*, 1159–1165. [CrossRef]
62. Boodhoo, K.V.K.; Flickinger, M.C.; Woodley, J.M.; Emanuelsson, E.A.C. Bioprocess Intensification: A Route to Efficient and Sustainable Biocatalytic Transformations for the Future. *Chem. Eng. Process. Process Intensif.* **2022**, *172*, 108793. [CrossRef]
63. Monck, C.; Elani, Y.; Ceroni, F. Cell-Free Protein Synthesis: Biomedical Applications and Future Perspectives. *Chem. Eng. Res. Des.* **2022**, *177*, 653–658. [CrossRef]
64. Banhazi, T.; Babinszky, L.; Halas, V.; Tscharke, M. Precision Livestock Farming: Precision Feeding Technologies and Sustainable Animal Production. *Int. J. Agric. Biol. Eng.* **2012**, *5*, 54–61. [CrossRef]
65. Pomar, C.; van Milgen, J.; Remus, A. Precision Livestock Feeding, Principle and Practice. In *Poultry and Pig Nutrition: Challenges of the 21st Century*; Hendriks, W.H., Verstegen, M.W.A., Babinszky, L., Eds.; Wageningen Academic Publishers: Wageningen, The Netherlands, 2019; ISBN 978-90-8686-333-4.
66. Zuidhof, M.J.; Fedorak, M.V.; Ouellette, C.A.; Wenger, I.I. Precision Feeding: Innovative Management of Broiler Breeder Feed Intake and Flock Uniformity. *Poult. Sci.* **2017**, *96*, 2254–2263. [CrossRef]
67. Li, G.; Zhao, Y.; Hailey, R.; Zhang, N.; Liang, Y.; Purswell, J.L. An Ultra-High Frequency Radio Frequency Identification System for Studying Individual Feeding and Drinking Behaviors of Group-Housed Broilers. *Animal* **2019**, *13*, 2060–2069. [CrossRef]

animals

MDPI

Article

Computer-Vision-Based Indexes for Analyzing Broiler Response to Rearing Environment: A Proof of Concept

Juliana Maria Massari [1], Daniella Jorge de Moura [1], Irenilza de Alencar Nääs [2,*], Danilo Florentino Pereira [3] and Tatiane Branco [1]

[1] College of Agricultural Engineering, State University of Campinas, 501 Candido Rondon Avenue, Campinas, São Paulo 13083-875, Brazil; jujumassari@hotmail.com (J.M.M.); djmoura@unicamp.br (D.J.d.M.); tatibranco91@gmail.com (T.B.)

[2] Graduate Program in Production Engineering, Universidade Paulista, 1212 Dr. Bacelar Street, São Paulo 04026-002, Brazil

[3] Department of Management, Development and Technology, School of Science and Engineering, São Paulo State University, 780 Domingos da Costa Lopes Avenue, Tupã, São Paulo 17602-496, Brazil; danilo.florentino@unesp.br

* Correspondence: irenilza.naas@docente.unip.br

Simple Summary: We tested two computer-vision-based indexes to analyze the rearing-environment enrichment on broiler movement as a function of comfort temperature and heat stress. The results indicated that the simultaneous application of cluster and unrest indexes could monitor the movement of the group of broilers under different environmental conditions. Future monitoring and alert systems based on computer vision should consider the complexity of the environment for detecting heat stress in broiler production.

Abstract: Computer-vision systems for herd detection and monitoring are increasingly present in precision livestock. This technology provides insights into how environmental variations affect the group's movement pattern. We hypothesize that the cluster and unrest indexes based on computer vision (CV) can simultaneously assess the movement variation of reared broilers under different environmental conditions. The present study is a proof of principle and was carried out with twenty broilers (commercial strain Cobb®), housed in a controlled-environment chamber. The birds were divided into two groups, one housed in an enriched environment and the control. Both groups were subjected to thermal comfort conditions and heat stress. Image analysis of individual or group behavior is the basis for generating animal-monitoring indexes, capable of creating real-time alert systems, predicting welfare, health, environment, and production status. The results obtained in the experiment in a controlled environment allowed the validation of the simultaneous application of cluster and unrest indexes by monitoring the movement of the group of broilers under different environmental conditions. Observational results also suggest that research in more significant proportions should be carried out to evaluate the potential positive impact of environmental enrichment in poultry production. The complexity of the environment is a factor to be considered in creating alert systems for detecting heat stress in broiler production. In large groups, birds' movement and grouping patterns may differ; therefore, the CV system and indices will need to be recalibrated.

Keywords: walking ability; animal welfare; animal behavior; image analysis; precision livestock

Citation: Massari, J.M.; de Moura, D.J.; de Alencar Nääs, I.; Pereira, D.F.; Branco, T. Computer-Vision-Based Indexes for Analyzing Broiler Response to Rearing Environment: A Proof of Concept. *Animals* **2022**, *12*, 846. https://doi.org/10.3390/ani12070846

Academic Editor: Janice Siegford

Received: 28 February 2022
Accepted: 26 March 2022
Published: 28 March 2022

1. Introduction

Prospects of future scenarios indicate that the world population will grow to 9.3 billion people in 2050, which requires a significant increase in food demand. Globally, chicken meat is expected to represent 41% of all animal-protein sources by 2030 [1]. In order to meet this strong demand for market growth, intensive production of broilers has prevailed. In most of these systems, the rearing environment restricts opportunities for species-specific behaviors, which are essential for good welfare [2]. Broilers are housed at a high density and with selected genetic characteristics for rapid growth [3]. However, the bone structure of the chicken did not follow this process of high development of the upper body part (breast), which triggers leg disorders and the consequent loss of mobility with increasing body weight [4].

The automatic detection of the activity level of groups of broilers makes it possible to identify deviations outside the expected patterns and generate real-time notification alerts to the producer, which allows a faster readjustment with benefits for the welfare of the animals [5]. Stimulating physical activity in birds prevents the occurrence of locomotor problems that impair wellbeing [6,7]. Previous studies indicate that enriched environments have the positive potential to stimulate and increase the activity level of broilers [4,8–12]. Physical activity strengthens the locomotor system, especially at the beginning of the growth phase [13]. In summary, environmental enrichment introduces improvements in existing production systems and considers which artifacts stimulate the behavioral activities inherent to the species, promoting improvements in biological function [14,15].

Computer vision (CV) applies mathematics and computer science to provide image-based automated process control [16]. CV allows continuous and real-time measurements during the flock production cycle in a fully automated, noninvasive way [17]. The data images were collected past steps to preprocessing, segmentation (region of interest), features extraction, and classification or regression [16]. Thus, the producer can monitor various biological processes and bioresponses related to animal welfare, health, feeding and drinking behaviors, and flock productivity [18–20]. We hypothesize that computer vision associated with movement indexes can monitor locomotor-health problems and prevalence in broiler flocks [5,21–24]. Studies based on proof of concept are present in animal production and evaluate the technical, practical, and financial feasibility of an idea or hypothesis [25,26]. Several studies have been developed to monitor locomotor-health [22,27–29] body-mass estimation [30,31]. The effect of environmental enrichment on broiler activity levels, gait assessment, locomotor problems, zootechnical performance, and behavior and wellbeing have been previously studied using video-image-processing techniques [32–35]. The computer-vision technology was also validated in a laboratory scale for automatic monitoring and gait-score classification [36]. It was also used to identify abnormal deviations in the activity level of commercial birds [5] and for evaluating the occupancy rate of laying hens in compartments with different levels of ammonia concentration [37]. Under heat-stress conditions, there is a significant decrease in growth rate, increased mortality, a compromised immune system, loss of meat quality, behavioral changes, and a decreased level of wellbeing [38,39]. The rapid diagnosis of animals in thermal discomfort is crucial to prevent the stress from being prolonged, preserving broiler performance, health, and welfare.

The continuous analysis of image processing obtained by video cameras allows for the generation of activity indexes that monitor the thermal state of broilers [40–42]. Most studies in the current literature involving heat stress were conducted from 21 days of age [43]. Animals change their behavioral pattern as a function of the rearing temperature, being close to each other when subjected to cold or spread out in the environment in the heat [42]. Livestock workers have routinely used these postural patterns to assess thermal comfort and adjust to environmental management settings [44]. Thus, observation of behavioral parameters is a noninvasive way of detecting heat stress [45]. Previously CV has been used for the generation of cluster and unrest indices, developed respectively by [23] and [24], and have been applied as indicators of thermal comfort in commercial poultry production. The results obtained indicated that the unrest index could detect the agitation of poultry under different thermal conditions, with a significant decrease in the movement of birds under heat stress [24].

On the other hand, the cluster index revealed a significant difference in the clustering behavior of birds under conditions of comfort and heat stress. In addition, it identified behavioral differences between the heavy rearing breeds [23]. The unrest index was used to measure the walking ability of broilers with different gait scores [46]. Both indices have the potential to develop a remote-monitoring system to accurately detect differences in the behavior of birds raised in floor pens bedded with wood shavings. The application of these indices has not yet been explored in the rearing of broilers in enriched environments.

This study is a proof of concept to assess the use of the cluster and unrest indexes simultaneously to test the sensitivity and viability in order to evaluate the movement of broilers under different conditions such as heat stress and pen enrichment.

2. Materials and Methods

2.1. Description of the Controlled-Environment Chamber

The controlled-temperature room has three compartments (C1, C2 and C3), measuring $1.6 \times 1.4 \times 3.0$ m^3. Only two compartments were used for the present study, and they were randomly selected. In each compartment, a manual tube-type feeder (Zatti® Model number 181,528, Zatti Industry and Commerce, Coronel Freitas, Santa Catarina, Brazil), an automatic pendulum-type drinker (CASP®, Model pendular drinking automatic 2003, CASP, Amparo, São Paulo, Brazil), and a temperature and humidity sensor were installed close to the animals' level, at a distance of 0.40 m from the floor. Each compartment had an air conditioner, two dehumidifiers, two heaters, a dimmable LED lamp to control light intensity (lx), and a video camera. Each compartment was accessed independently through a door (0.7 m wide $\times$ 2 m high). The computers responsible for managing the experimental units were installed in the support room (climate control and video recording). Figure 1 [47] shows the schematic with the respective positioning of all the equipment used in the controlled-temperature room and the technical-support room during the experimental period.

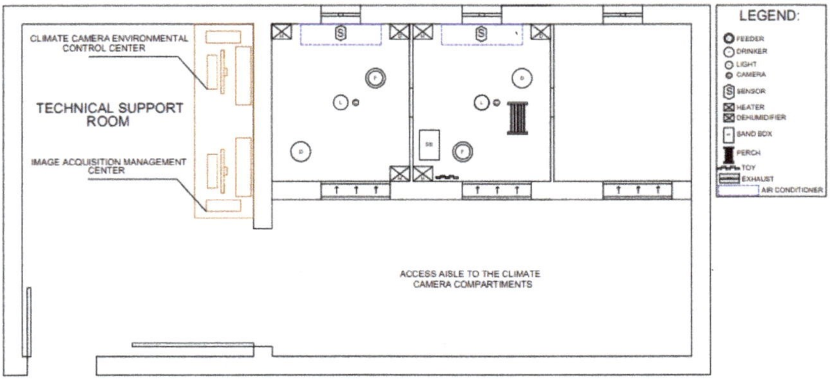

Figure 1. Plan view of the environmental chamber (adapted from [47]). Reprinted/adapted with permission from Ref. [47]. 2022, Daniella Jorge de Moura.

The environmental-control center manages each compartment of the controlled-environment room using software developed in the Delphi programming language (version 6.0, Borland Software Co., Austin, TX, USA). The software allows measuring, processing, controlling, and recording continuously collected data. This system allows the user to check temperature, humidity, light intensity, and air renewal rates in real time. According to the established temperature and humidity values, the equipment is automatically activated (on and off). Relative humidity was programmed to remain at 60% continuously. Only the air temperature varied during the experimental period, with operating values recommended by the [48] of 23 °C for thermal comfort and 31 °C for heat-stress treatment. The particularities of the environmental-control system, equipment, operating limits, stability, and validation with broilers are presented in [47]. The steps of inputs, data collection and processing, experimental treatments, and outputs are represented in Figure 2 and explained in the sequence.

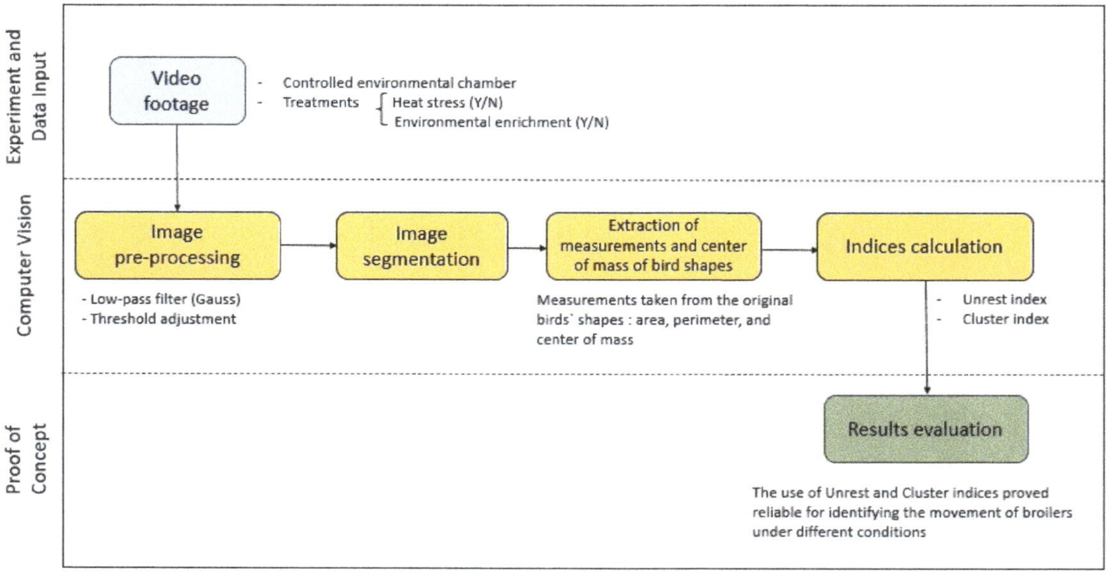

Figure 2. Diagram of data input and output of the tests.

2.2. Image Acquisition

For the animal-behavior monitoring data and further analysis, surveillance cameras (Intelbras® VMD 3120 IR, Intelbras Corporation, São José, Santa Catarina, Brazil) with a resolution of 976 × 496 (H × V) and automatic activation of the infrared device in cases of low light were installed on the ceiling of the geometric center of each compartment. The two validation tests used recorded video images between noon and 18:00 h. Video recordings were automatically stored on an NVR video recorder (Intelbras® Multi HD Serie 1000, 1080p, Intelbras Corporation, São José, Santa Catarina, Brazil). Figure 3 shows the areas observed for the unenriched (a) and enriched (b) compartments.

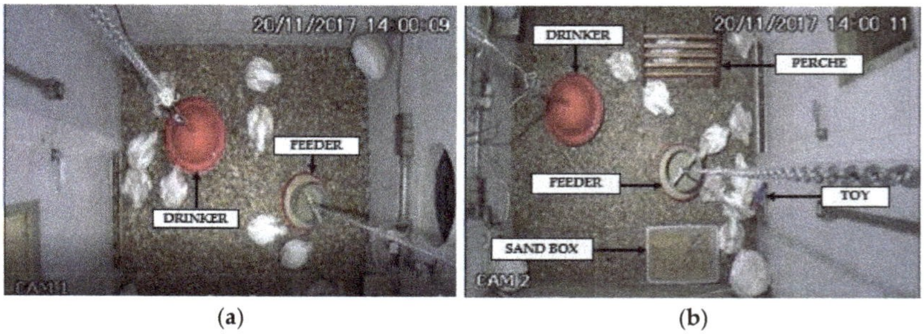

Figure 3. Top view of the unenriched (**a**) and enriched (**b**) compartment through the cameras installed on the ceiling of each compartment of the climate chamber.

2.3. Birds and Husbandry

A total of thirty-day-old mixed-sex chicks of the Cobb® strain were obtained from a commercial farm. Twenty chicks with similar weights and the same distribution of males and females were selected in two treatments, each containing ten birds (without environmental enrichment and with environmental enrichment). The compartments and animals were randomly assigned to assign the treatments on the first day of housing in the controlled-environment chamber.

Both compartments were kept without environmental enrichment during the first three days of adaptation in the climatic chamber. The compartment selected as "enriched" was provided with colored plastic rings suspended by a string, a plastic box containing fine sand, and a wooden perch. According to previous literature, the enrichment was selected to positively affect the birds' natural behaviors (perching, pecking, and dust bathing) [13,49,50].

A bell drinker and feeder were placed in each compartment. Water and commercial feed, based on corn and soybean meal, were provided ad libitum throughout the rearing period. The supply of commercial feed and water to the birds was ad libitum throughout the experimental period and followed the nutritional recommendations of the breeder's manual [48]. Once a day, the offered ration was weighed and manually inserted into the tube feeder. The automatic water-supply system used a tubular drinking fountain with height adjustment. The floor was covered with shavings bedding (0.05 m). We adopted the breeding-company-recommended period of light (24 h of light until the seventh day and increasing 1 h of darkness every two days). On the 14th day of growth, the birds remained exposed to 20 h of light and 4 h of darkness from 21:00–01:00 until the end of the experiment (42nd day of growth). We also adopted the breeding manual, so the broilers were kept in thermoneutrality conditions during the first to the twentieth day of growth [48].

The acquisition of video images was performed automatically for seven consecutive hours from noon to 18:00 h for the two consecutive days of analysis. According to a previous study [5], broilers' activity patterns were similar for three weeks throughout the day. Such assumption allowed us to validate the analysis of broiler movement through the cluster and unrest indexes in two days for experimental conditions in the controlled-environment room, characterizing the present study as a proof of concept. The age of 21 days is when the heat stress starts to impair productive performance (decrease in feed consumption and weight gain) and negatively challenge animal metabolism and immunity [43,51,52]. Our experimental tests were precisely at the age of 21 and 22 days under thermoneutrality and heat-stress conditions, respectively, for both treatments. Heat stress was tested for one day, with the chickens at 22 days of age, with the birds being kept in thermal comfort during the previous housing days. Figure 4 illustrates the diagram of activities used in data collection for proof-of-concept validation.

The relative humidity remained within what was recommended in the breeders' manual [48] during the experiment. A previous commissioning study of the controlled-environment-room operation [47] allowed one hour to reach the heat-stress temperature condition (31 °C) and maintain the system. For this reason, the control setup started at noon.

2.4. Video Analysis

The efficiency of two comfort indexes based on group behavior was verified. The calculation of these indexes is based on information extracted from images recorded through image-analysis techniques. This proof of concept evaluated the extraction of information and the calculation of indexes by a computer-vision system.

Videos were analyzed at the frequency of one frame per second (fps). Considering that there was no effect of the compartments, we used a completely randomized design in a split-plot scheme in time, in which we tested two factors: (1) temperature (comfort or heat stress), as the main factor; and (2) environmental enrichment (present or absent), as the secondary factor. The seven hours of recording analyzed were divided into 14 blocks of time (30 min each block, which corresponds to the analysis of 25,200 frames per condition (temperature vs. environment), totaling 100,800 frames in the experimental period.

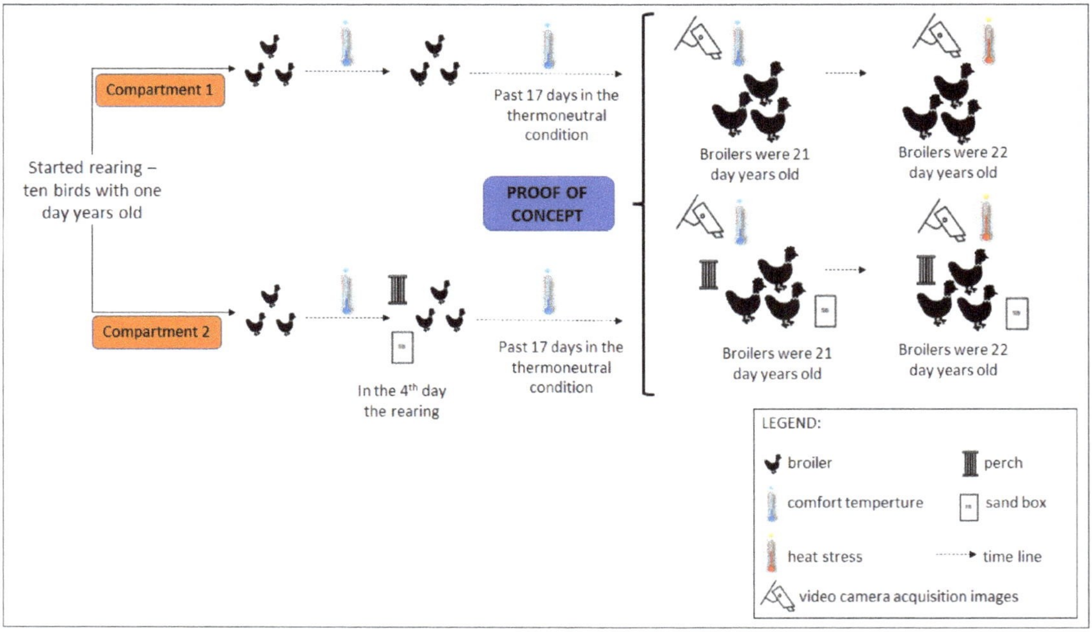

Figure 4. Diagram of the methodology used for proof-of-concept tests.

The images were processed frame by frame, initially using low-pass filters to smooth out image noises such as feathers on the bedding and wood shavings on the birds. After segmentation, mathematical morphology techniques were applied to fill holes and exclude the remaining noise. The group behavior of chickens was measured using the cluster index [23] and the unrest index [24], described in Equations (1) and (2).

$$\text{Cluster Index}_{\text{i}} = \frac{2 \times \overline{A} \times \sqrt{h^2 + w^2}}{\overline{P} \times \overline{D} \times n_A} - 1 \quad (1)$$

where Cluster $\text{Index}_{(\text{i})}$ is the cluster index of the birds observed in the ith frame of the video; $\overline{A}$ and $\overline{P}$ are the average area and perimeter (in pixels) of the shapes observed in the frame, respectively; $\overline{D}$ is the average distance between the centers of mass of the shapes in the scene; n_A are the number of clusters; and h and w correspond to the height and width (in pixels) of the cropped image.

$$\text{Unrest Index}_{(\text{i,i}-1)} = k.\max\left\{\text{dH}\left(\text{F}_{(\text{i})}, \text{F}_{(\text{i}-1)}\right), \text{dH}\left(\text{F}_{(\text{i}-1)}, \text{F}_{(\text{i})}\right)\right\} \quad (2)$$

where Unrest $\text{Index}_{(\text{i, i}-1)}$ is the unrest index (cm) of the birds between two frames recorded with 1 (one) second difference; i is the position of the frame in the video; $\text{F}_{(\text{i})}$ is the current frame; $\text{F}_{(\text{i}-1)}$ is the previous frame; dH is the Hausdorff distance [53] between birds from one frame to another; and k is the proportionality factor calculated by Equation (3).

$$k = \frac{2H \tan(\alpha/2)}{w} \quad (3)$$

where k is the proportionality factor; H is the height (cm) of the installed camera concerning the floor; α is the opening angle of the camera lens; and w is the length (pixels) of the CCD sensor, which corresponds to the length of the largest measurement of the frame captured by the camera. The video capture rate was 30 fps, but a frequency of 1 fps was adopted as the most adequate for image analysis, considering the birds' movement speed.

Cluster and unrest indexes were calculated frame by frame. In this way, the values obtained for each plot correspond to an average referring to 1800 images. The data were explored by graphs of the indexes calculated in the simultaneous application observation time, verifying possible interactions and differences in the crowding and unrest behaviors between the enrichment and temperature treatments evaluated. We applied the ANOVA with repeated measures, followed by the Tukey mean test to confirm the differences in the birds' crowding and movement behavior between the evaluated environmental treatments.

3. Results and Discussion

Figure 5 corresponds to the results obtained for the cluster index (crowding) from noon to 18:00 in the group of broilers housed in an enriched and nonenriched environment, subjected to thermal conditions of neutrality and heat stress. Each point on the graph represents a repetition of a subdivision plot in time (total 14). Each plot corresponds to the analysis average of 1800 frames/video for generating the cluster index for the treatments.

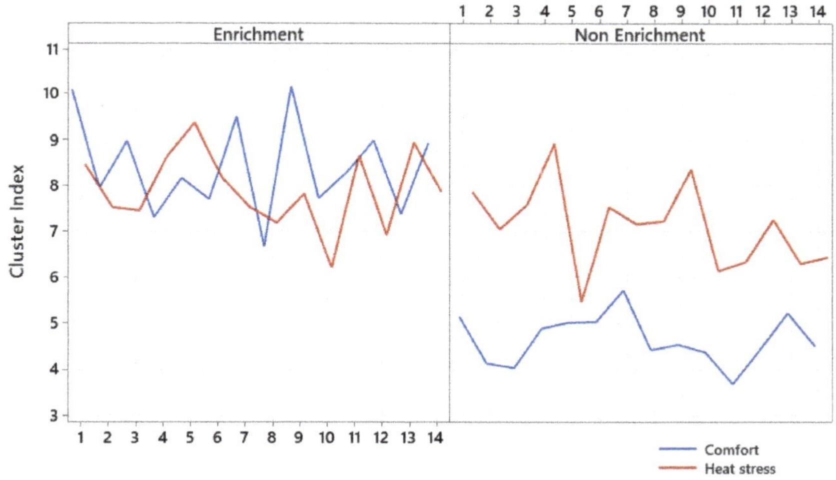

Figure 5. Cluster index of the group of broilers housed in enriched and nonenriched environments under the effect of comfort and heat stress.

It can be seen from Figure 5 that the cluster index detected in the enriched environment is similar between the conditions of thermoneutrality and heat stress. However, the highest peaks occurred in thermoneutrality and the lowest in heat stress. This observation of trend analysis means that the ambient temperature above the comfort limit was not enough to change the crowding pattern of broilers aged 21 days reared in enriched environments. Therefore, the isolated analysis of this index cannot be considered an indicator of heat stress for broilers raised in enriched environments. We observed significantly lower crowding rates in the comfort temperature and nonenriched environment than the enriched-compartment results.

Figure 6 illustrates the unrest index for the treatments from noon to 18:00 h. Each point of the graph represents a repetition of the subdivision plot in time (total 14). Each plot corresponds to the average of 1800 frames/video analysis for the generation of the unrest index of the treatments.

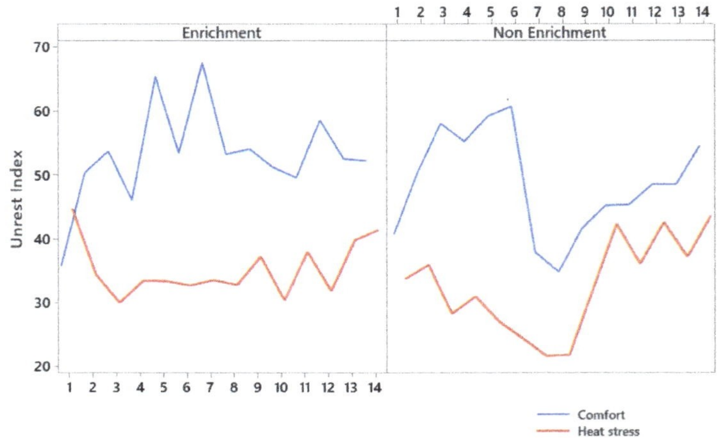

Figure 6. Unrest index of the group of broilers housed in enriched and nonenriched environments under the effect of comfort and heat stress.

During the evaluated period, a tendency for the unrest index to be higher under thermoneutrality conditions is observed, regardless of the environmental enrichment. Furthermore, heat stress reduces the movement of animals, with more significant losses for the nonenriched treatment. Table 1 shows the differences observed in the behavior of gathering and movement of the birds, where it is observed that the environmental enrichment promoted more movement of the birds both under conditions of comfort (52.74 vs. 48.71) and thermal stress (35.38 vs. 32.81), which confirm previous graphical analyses. However, heat stress is a limiting factor for the movement of animals, reducing the positive potential of the presence of environmental enrichment. Environmental enrichment provided higher crowding rates both in comfort (8.42 vs. 4.66) and in heat stress (7.91 vs. 7.11), and in an enriched environment, birds under heat stress crowded more than birds raised in nonenriched environments (4.66 vs. 7.11).

Table 1. Results of the Tukey test ($p < 0.05$) for the cluster and unrest indexes in the combined conditions of temperature and environmental enrichment found in the proof-of-concept experiment.

Indexes	Temperature	Environment	
		Enriched	**Nonenriched**
Cluster	Comfort	8.42	4.66
	Heat stress	7.91	7.11
Unrest	Comfort	52.74	48.71
	Heat stress	35.38	32.81

Environmental enrichment increased broilers' unrest (movement) index from fast and slow-growing strains under thermoneutrality conditions, confirming our findings [32]. Exposure to high temperatures above the thermal comfort zone is challenging for birds housed in complex environments. Broilers raised in enriched environments from 1-day-old exposed to stress conditions at 22 days of age (heat, noise, and containment in a crate) showed heat stress as the worst adverse condition [54].

Environmental enrichment benefits, especially perches and litter boxes, have been extensively studied [2,4,7,9–11,50]. In the present study, we noted behavioral changes in the group compatible with these benefits, suggesting that the proposed computer vision based on cluster and unrest indexes can be safely used for these assessments.

Figure 5 shows the crowding behavior of broilers housed in enriched and nonenriched environments at the different temperatures tested. Results of the present study indicated that complex environments favored the crowding of the group of broilers under heat stress. Visually reviewing the videos, we observed that birds under heat stress conditions clustered around enrichment objects, indicating that environmental enrichment can minimize the negative effect of heat stress on birds. Figure 7 illustrates the frames for comfort (a and c) and heat stress (b and d) conditions in enriched (a and b) and nonenriched (c and d) environments.

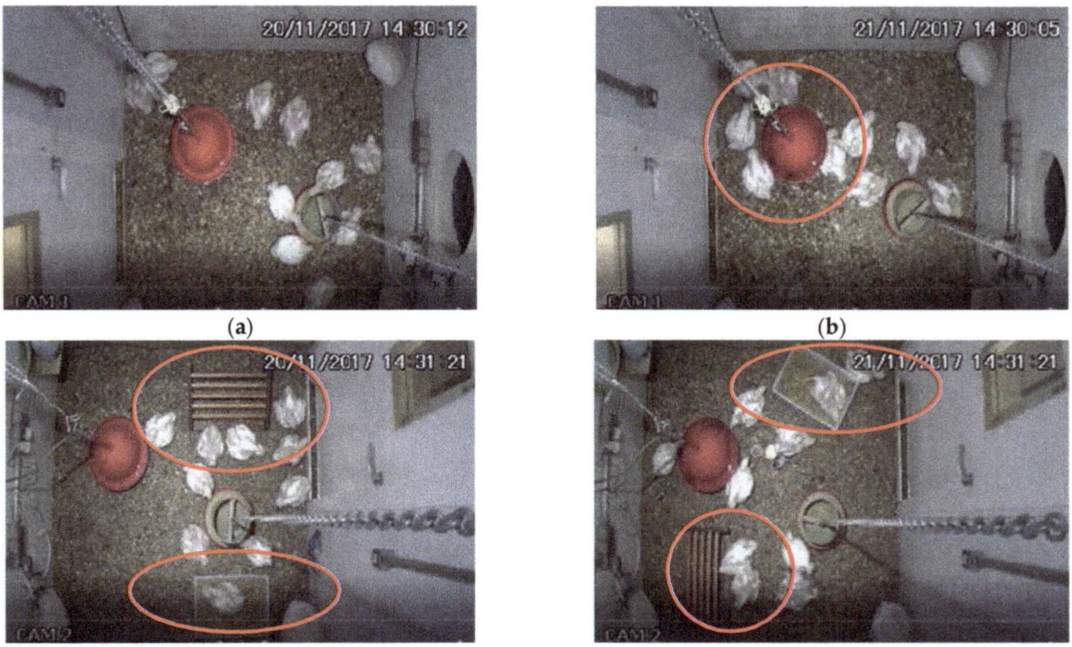

(a) (b) (c) (d)

Figure 7. Frames evaluated are (**a**) nonenriched environment in comfort, (**b**) nonenriched environment under heat stress, (**c**) environment enriched in comfort, and (**d**) enriched environment under heat stress.

When analyzing the positioning of the animals in the video images, it was observed that broilers have different distribution patterns depending on the complexity of the environment (enriched versus nonenriched). Note that the birds are better distributed throughout the compartment for the environment without environmental enrichment in the comfort treatment (Figure 7a). However, in the heat-stress condition (Figure 7b), birds crowded near the drinker to benefit from the microclimate close to the water, which resulted in higher crowding rates [23,55]. The behaviors of remaining seated birds—increased water consumption, spreading wings, increased respiratory rate, and panting—are favored to dissipate excess heat [55–58].

In our study and previous work, bird distribution inside the pen appears to be highly related to the location of food and water [59]. Environmental enrichment also alters the distribution pattern of birds, with a higher prevalence of agglomeration close to enrichment objects, both in small-group experiments [32,60,61] and at a commercial scale [2,8,62]. Cornetto and Estevez [60] observed that the birds were forced to occupy the central region earlier in groups of larger sizes than for smaller group sizes. Slow-growing broilers used environmental-enrichment objects more frequently when compared to fast-growing broilers [32]. These findings reinforce the differences among housed flocks and the importance of recalibrating the CV system to each housing situation (group size, strains, age, and housing conditions).

The CV could simultaneously apply the Cluster and Unrest indexes to monitor the movement of the group of broilers under different environmental conditions, indicating the possible differences in the environmental conditions. The authors suggest that more research should be conducted to evaluate the potential positive impact of environmental enrichment in poultry production. The complexity of the environment is a factor to be considered in creating alert systems for detecting heat stress in broiler production.

4. Conclusions

The cluster and unrest indexes calculated from videos analyzed by computer-vision techniques allowed us to simultaneously evaluate the movement of broilers raised in different environments and detect variations that allowed us to estimate the level of wellbeing. We recommend that the indexes be used to evaluate the movement and agglomeration of broiler flocks in environments with different enrichment levels to evaluate the improvement of wellbeing. In large groups, birds' movement and grouping patterns may differ; therefore, the CV system and indices will need to be recalibrated. The use of CV to assist with monitoring can assist caregivers during the rearing of broiler chickens.

Author Contributions: Conceptualization, J.M.M. and D.J.d.M.; methodology, J.M.M., D.J.d.M. and T.B.; software, D.F.P.; validation, D.F.P. and T.B.; formal analysis, D.F.P.; data curation, D.F.P.; writing—original draft preparation, J.M.M., D.J.d.M. and D.F.P.; writing—review and editing, I.d.A.N., D.F.P. and D.J.d.M.; visualization, I.d.A.N., D.F.P. and D.J.d.M.; supervision, D.J.d.M., D.F.P. and I.d.A.N. All authors have read and agreed to the published version of the manuscript.

Funding: We acknowledge CNPq for the grant number 308177/2021-7 and 304085/2021-9.

Institutional Review Board Statement: This research was conducted according to the guidelines of the Declaration of Helsinki, and approved by the Ethics Committee on the use of animals (CEUA) of the State University of Campinas (UNICAMP), under protocol 4716-1/2017. The study was conducted in a climate chamber, located in the experimental area of the Thermal Comfort Laboratory of the College of Agricultural Engineering, State University of Campinas (UNICAMP), São Paulo, Brazil.

Informed Consent Statement: License granted by Elsevier, number 5273660324060, by 21 March 2022.

Data Availability Statement: Data will be available upon request.

Acknowledgments: The authors are grateful to the College of Agricultural Engineering (FEAGRI/UNICAMP) and State University of Campinas (UNICAMP). The assistance with commissioning tests provided by the research team of Thermal Comfort Lab undergraduate students was greatly appreciated.

Conflicts of Interest: The authors declare no conflict of interest.

References

1. Pocketbook, F.S. *World Food and Agriculture*; FAO: Rome Italy, 2015. Available online: https://agris.fao.org/agris-search/search.do?recordID=XF2016019183 (accessed on 5 May 2020).
2. Bergmann, S.; Schwarzer, A.; Wilutzky, K.; Louton, H.; Bachmeier, J.; Schmidt, P.; Erhard, M.; Rauch, E. Behavior as welfare indicator for the rearing of broilers in an enriched husbandry environment—A field study. *J. Vet. Behav.* **2017**, *19*, 90–101. [CrossRef]
3. Zuidhof, M.J.; Schneider, B.L.; Carney, V.L.; Korver, D.R.; Robinson, F.E. Growth, efficiency, and yield of commercial broilers from 1957, 1978, and 2005. *Poult. Sci.* **2014**, *93*, 2970–2982. [CrossRef] [PubMed]
4. Meyer, M.M.; Johnson, A.K.; Bobeck, E.A. A novel environmental enrichment device increased physical activity and walking distance in broilers. *Poult. Sci.* **2020**, *99*, 48–60. [CrossRef] [PubMed]
5. Kristensen, H.H.; Cornou, C. Automatic detection of deviations in activity levels in groups of broiler chickens—A pilot study. *Biosyst. Eng.* **2011**, *109*, 369–376. [CrossRef]
6. Kestin, S.C.; Knowles, T.G.; Tinch, A.E.; Gregory, N.G. Prevalence of leg weakness in broiler chickens and its relationship with genotype. *Vet. Rec.* **1992**, *131*, 190–194. [CrossRef]
7. Reiter, K.; Bessei, W. Effect of locomotor activity on leg disorder in fattening chicken. *Berl. Munch. Tierarztl. Wochenschr.* **2009**, *122*, 264–270.
8. De Jong, I.C.; Gunnink, H. Effects of a commercial broiler enrichment programme with or without natural light on behaviour and other welfare indicators. *Animal* **2019**, *13*, 384–391. [CrossRef]

9. Bach, M.H.; Tahamtani, F.M.; Pedersen, I.J.; Riber, A.B. Effects of environmental complexity on behaviour in fast-growing broiler chickens. *Appl. Anim. Behav. Sci* **2019**, *219*, 104840. [CrossRef]
10. Kaukonen, E.; Norring, M.; Valros, A. Perches and elevated platforms in commercial broiler farms: Use and effect on walking ability, incidence of tibial dyschondroplasia and bone mineral content. *Animal* **2017**, *11*, 864–871. [CrossRef]
11. Vasdal, G.; Vas, J.; Newberry, R.C.; Moe, R.O. Effects of environmental enrichment on activity and lameness in commercial broiler production. *J. Appl. Anim. Welf. Sci.* **2019**, *22*, 197–205. [CrossRef]
12. Ventura, B.A.; Siewerdt, F.; Estevez, I. Effects of barrier perches and density on broiler leg health, fear, and performance. *Poult. Sci.* **2010**, *89*, 1574–1583. [CrossRef] [PubMed]
13. Bizeray, D.; Estevez, I.; Leterrier, C.; Faure, J.M. Effects of increasing environmental complexity on the physical activity of broiler chickens. *Appl. Anim. Behav. Sci.* **2002**, *79*, 27–41. [CrossRef]
14. Newberry, R.C. Environmental enrichment: Increasing the biological relevance of captive environments. *Appl. Anim. Behav. Sci.* **1995**, *44*, 229–243. [CrossRef]
15. Van de Weerd, H.A.; Day, J.E.L. A review of environmental enrichment for pigs housed in intensive housing systems. *Appl. Anim. Behav. Sci* **2009**, *116*, 1–20. [CrossRef]
16. Okinda, C.; Nyalala, I.; Korohou, T.; Okinda, C.; Wang, J.; Achieng, T.; Wamalwa, P.; Mang, T.; Shen, M. A review on computer vision systems in monitoring of poultry: A welfare perspective. *Artif. Intell. Agric.* **2020**, *4*, 184–208. [CrossRef]
17. Dawkins, M.S.; Lee, H.J.; Waitt, C.D.; Roberts, S.J. Optical flow patterns in broiler chicken flocks as automated measures of behaviour and gait. *Appl. Anim. Behav. Sci.* **2009**, *119*, 203–209. [CrossRef]
18. Berckmans, D. General introduction to precision livestock farming. *Anim. Front.* **2017**, *7*, 6–11. [CrossRef]
19. Li, G.; Zhao, Y.; Purswell, J.L.; Du, Q.; Chesser, G.D.; Lowe, J.W. Analysis of feeding and drinking behaviors of group-reared broilers via image processing. *Comput. Electron. Agric.* **2020**, *175*, 105596. [CrossRef]
20. Wathes, C.M.; Kristensen, H.H.; Aerts, J.M.; Berckmans, D. Is precision livestock farming an engineer's daydream or nightmare, an animal's friend or foe, and a farmer's panacea or pitfall? *Comput. Electron. Agric.* **2008**, *64*, 2–10. [CrossRef]
21. Kashiha, M.; Pluk, A.; Bahr, C.; Vranken, E.; Berckmans, D. Development of an early warning system for a broiler house using computer vision. *Biosyst. Eng.* **2013**, *116*, 36–45. [CrossRef]
22. Van Hertem, T.; Norton, T.; Berckmans, D.; Vranken, E. Predicting broiler gait scores from activity monitoring and flock data. *Biosyst. Eng.* **2018**, *173*, 93–102. [CrossRef]
23. Pereira, D.F.; Lopes, F.A.A.; Filho, L.R.A.G.; Salgado, D.D.A.; Neto, M.M. Cluster index for estimating thermal poultry stress (gallus gallus domesticus). *Comput. Electron. Agric.* **2020**, *177*, 105704. [CrossRef]
24. Del Valle, J.E.; Pereira, D.F.; Mollo Neto, M.; Gabriel Filho, L.R.A.; Salgado, D.D.A. Unrest index for estimating thermal comfort of poultry birds (Gallus gallus domesticus) using computer vision techniques. *Biosyst. Eng.* **2021**, *206*, 123–134. [CrossRef]
25. Bloch, V.; Barchilon, N.; Halachmi, I.; Druyan, S. Automatic broiler temperature measuring by thermal camera. *Biosyt. Eng.* **2019**, *199*, 127–134. [CrossRef]
26. Hernandez-Patlan, D.; Solis-Cruz, B.; Adhikari, B.; Pontin, K.P.; Latorre, J.D.; Baxter, M.F.A.; Hernandez-Veloso, X.; Merino-Guzman, R.; Méndez-Albores, A.; Kwon, Y.M.; et al. Evaluation of the antimicrobial and intestinal integrity properties of boric acid in broiler chickens infected with *Salmonella enteritidis*: Proof of concept. *Res. Vet. Sci.* **2019**, *123*, 7–13. [CrossRef]
27. Aydin, A. Development of an early detection system for lameness of broilers using computer vision. *Comput. Electron. Agric.* **2017**, *136*, 140–146. [CrossRef]
28. Aydin, A. Using 3D vision camera system to automatically assess the level of inactivity in broiler chickens. *Comput. Electron. Agric.* **2017**, *135*, 4–10. [CrossRef]
29. Silvera, A.M.; Knowles, T.G.; Butterworth, A.; Berckmans, D.; Vranken, E.; Blokhuis, H.J. Lameness assessment with automatic monitoring of activity in commercial broiler flocks. *Poult. Sci.* **2007**, *7*, 2013–2017. [CrossRef]
30. Mortensen, A.K.; Lisouski, P.; Ahrendt, P. Weight prediction of broiler chickens using 3D computer vision. *Comput. Electron. Agric.* **2016**, *123*, 319–326. [CrossRef]
31. De Wet, L.; Vranken, E.; Chedad, A.; Aerts, J.M.; Ceunen, J.; Berckmans, D. Computer-assisted image analysis to quantify daily growth rates of broiler chickens. *Br. Poult. Sci* **2003**, *44*, 524–532. [CrossRef]
32. de Jong, I.C.; Blaauw, X.E.; van der Eijk, J.A.J.; da Silva, C.S.; Van Krimpen, M.M.; Molenaar, R.; Van Den Brand, H. Providing environmental enrichments affects activity and performance, but not leg health in fast-and slower-growing broiler chickens. *Appl. Anim. Behav. Sci* **2021**, *241*, 105375. [CrossRef]
33. Ross, L.; Wagner, B.K.; Cressman, M.D.; Cramer, M.C.; Pairis-Garcia, M.D. Short Communication: Investigating woody breast disease and broiler chicken activity using an automated tracking software program. *Appl. Anim. Sci.* **2020**, *36*, 447–453. [CrossRef]
34. Silva, M.I.L.; Paz, I.C.L.A.; Chaves, G.H.C.; Almeida, I.C.L.; Ouros, C.C.; de Souza, S.R.L.; Milbradt, E.L.; Caldara, F.R.; Satin, A.J.G.; Costa, G.A.; et al. Behaviour and animal welfare indicators of broiler chickens housed in an enriched environment. *PLoS ONE* **2021**, *16*, e0256963.
35. Dawson, L.C.; Widowski, T.M.; Liu, Z.; Edwards, A.M.; Torrey, S. In pursuit of a better broiler: A comparison of the inactivity, behavior, and enrichment use of fast- and slower growing broiler chickens. *Poult. Sci.* **2021**, *100*, 101451. [CrossRef] [PubMed]
36. Aydin, A.; Cangar, O.; Ozcan, S.E.; Bahr, C.; Berckmans, D. Application of a fully automatic analysis tool to assess the activity of broiler chickens with different gait scores. *Comput. Electron. Agric.* **2010**, *73*, 194–199. [CrossRef]

37. Kashiha, M.A.; Green, A.R.; Sales, T.G.; Bahr, C.; Berckmans, D.; Gat, R.S. Performance of an image analysis processing system for hen tracking in an environmental preference chamber. *Poult. Sci.* **2014**, *93*, 2439–2448. [CrossRef] [PubMed]
38. Shakeri, M.; Cottrell, J.J.; Wilkinson, S.; Le, H.H.; Suleria, H.A.R.; Warner, R.D.; Dunshea, F.R. Growth performance and characterization of meat quality of broiler chickens supplemented with betaine and antioxidants under cyclic heat stress. *Antioxidants* **2019**, *8*, 336. [CrossRef]
39. Slawinska, A.; Zampiga, M.; Sirri, F.; Meluzzi, A.; Bertocchi, M.; Tavaniello, S.; Maiorano, G. Impact of galactooligosaccharides delivered in ovo on mitigating negative effects of heat stress on performance and welfare of broilers. *Poult. Sci.* **2020**, *99*, 407–415. [CrossRef]
40. Gates, R.S.; Xin, H. Comparative Analysis of Measurement Techniques of Feeding Behavior of Individual. In Proceedings of the ASAE Metting Presentation, Sacramento, CA, USA, 30 July–1 August 2001; ASABE: St.Joseph, MO, USA, 2001.
41. Bloemen, H.; Aerts, J.M.; Berckmans, D.; Goedseels, V. Image analysis to measure activity index of animals. *Equine Vet. J. Suppl.* **1997**, *23*, 16–19. [CrossRef]
42. Youssef, A.; Exadaktylos, V.; Berckmans, D.A. Towards real-time control of chicken activity in a ventilated chamber. *Biosyst. Eng.* **2015**, *135*, 31–43. [CrossRef]
43. Andretta, I.; Kipper, M.; Schirmann, G.D.; Franceschina, C.S.; Ribeiro, A.M.L. Modeling the performance of broilers under heat stress. *Poult. Sci.* **2021**, *100*, 101338. [CrossRef] [PubMed]
44. Xin, H. Assessing swine thermal comfort by image analysis of postural behaviors. *J. Anim. Sci.* **1999**, *77*, 1–9. [CrossRef] [PubMed]
45. Hoffmann, G.; Herbut, P.; Pinto, S.; Heinicke, J.; Kuhla, B.; Amon, T. Animal-related, non-invasive indicators for determining heat stress in dairy cows. *Biosyst. Eng.* **2020**, *199*, 83–96. [CrossRef]
46. Pereira, D.F.; Nääs, I.D.A.; Lima, N.D.D.S. Movement Analysis to Associate Broiler Walking Ability with Gait Scoring. *AgriEngineering* **2021**, *3*, 394–402. [CrossRef]
47. Maia, A.P.A.; Moura, D.J.; Green, A.R.; Silva, W.T.; Sarubbi, J.; Massari, J.M.; Barbosa, L.V.S. Design and testing of a novel environmental preference chamber. *Comput. Electron. Agric.* **2019**, *157*, 23–37. [CrossRef]
48. Coob—Vantress Brasil, Ltd.a. Manual de Manejo de Frangos de Corte. 2013. Available online: https://www.cobb-vantress.com/assets/Cobb-Files/df5655a7e9/Broiler-Guide-2019-POR-WEB.pdf (accessed on 5 May 2020).
49. Nicol, C.J. Effects of environmental enrichment and gentle handling on behaviour and fear responses of transported broilers. *Appl. Anim. Behav. Sci.* **1992**, *33*, 367–380. [CrossRef]
50. Miller, K.A.; Mench, J.A. The differential effects of four types of environmental enrichment on the activity budgets, fearfulness, and social proximity preference of Japanese quail. *Appl. Anim. Behav. Sci.* **2005**, *95*, 169–187. [CrossRef]
51. Hu, H.; Bai, X.; Xu, K.; Zhang, C.; Chen, L. Effect of phloretin on growth performance, serum biochemical parameters and antioxidant profile in heat-stressed broilers. *Poult. Sci.* **2021**, *100*, 101217. [CrossRef]
52. Siddigui, S.H.; Krang, D.; Park, J.; Khan, M.; Belal, S.A.; Shin, D.; Shim, K. Altered relationship between gluconeogenesis and immunity in broilers exposed to heat stress for different durations. *Poult. Sci.* **2021**, *100*, 101274. [CrossRef]
53. Zhao, C.; Shi, W.; Deng, Y. A new Hausdorff distance for image matching. *Patt. Recog. Lett.* **2005**, *26*, 581–586. [CrossRef]
54. Altan, O.; Seremet, C.; Bayraktar, Ö. The effects of early environmental enrichment on performance, fear and physiological responses to acute stress of broiler. *Arch. Fur. Geflugelkd.* **2013**, *1*, 23–28.
55. Mack, L.A.; Felver-Gant, J.N.; Dennis, R.L.; Cheng, H.W. Genetic variations alter production and behavioral responses following heat stress in 2 strains of laying hens. *Poult. Sci.* **2013**, *92*, 285–294. [CrossRef] [PubMed]
56. Gowe, R.S.; Fairfull, R.W. Breeding for Resistance to Heat Stress. In *Poultry Production Hot Climate*; Daghir, N.J., Ed.; CABI: Wallingford, UK, 2008; pp. 13–29.
57. Sohail, M.U.; IJAZ, A.; Younus, M.; Shabbir, M.Z.; Kamran, Z.; Ahmad, S.; Anwar, H.; Yousaf, M.S.; Ashraf, K.; Shahzad, A.H.; et al. Effect of supplementation of mannan oligosaccharide and probiotic on growth performance, relative weights of viscera, and population of selected intestinal bacteria in cyclic heat-stressed broilers. *J. Appl. Poult. Res.* **2013**, *22*, 485–491. [CrossRef]
58. Pereira, D.F.; Nääs, A.; Salgado, D.A.; Gaspar, C.R.; Bighi, C.A.; Penha, N.L.J. Correlations among behavior, performance and environment in broiler breeders using multivariate analysis. *Rev. Bras. Cienc. Avic.* **2007**, *9*, 207–213. [CrossRef]
59. Mench, J.A.; Keeling, L.J. The social behaviour of domestic birds. In *Social Behaviour in Farm Animals*; Keeling, L.J., Gonyou, H.W., Eds.; CAB International: Wallingford, UK, 2001; pp. 177–210.
60. Cornetto, T.I. Estevez. Influence of vertical panels on use of space by domestic fowl. *Appl. Anim. Behav. Sci.* **2001**, *71*, 141–153. [CrossRef]
61. Cornetto, T.I. Estevez. Behavior of the domestic fowl in the presence of vertical panels. *Poult. Sci.* **2001**, *80*, 1455–1462. [CrossRef]
62. Baxter, M.; Bailie, C.L.; O'Connell, N.E. Evaluation of dustbathing susbtrate and straw bales as environmental enrichments in comercial broiler housing. *Appl. Anim. Behav. Sci.* **2018**, *200*, 78–85. [CrossRef]

MDPI
St. Alban-Anlage 66
4052 Basel
Switzerland
Tel. +41 61 683 77 34
Fax +41 61 302 89 18
www.mdpi.com

Animals Editorial Office
E-mail: animals@mdpi.com
www.mdpi.com/journal/animals

MDPI

www.ingramcontent.com/pod-product-compliance
Lightning Source LLC
Chambersburg PA
CBHW061616210926
51850CB00009B/2342
* 9 7 8 3 0 3 6 5 4 8 7 9 1 *